Making Maths Easy For All Students

Lower Level up to GCSE grade C

BRYCE RICHARD THOMAS

First published 1988 by:
Popplewells Publishing Company Ltd.,
Swiss Valley,
Llanelli,
Dyfed.
SA14 8LZ.

British Library Cataloguing-in-Publication Data.

Thomas, Bryce Richard.
Making Maths Easy.
1. Mathematics — For Schools.
I. Title
510.

ISBN 0-9513934-0-5.

Typeset and printed in Great Britain by John Penry Press, Swansea

Also by Bryce Richard Thomas

Non-Fiction: *Making Maths Easy: To Gain a Higher Level. Higher level up to GCSE Grade A.*

Fiction: *Time to Kill.*

To June
and Anna
Both are instrumental to the publication of this book.

Introduction

MAKING MATHEMATICS EASY

Aim.
The aim of this publication is to provide any student struggling or not, with the means to gaining AT LEAST the first 50% in his or her maths exam.
In particular, it is to aid those pupils who have difficulty in the subject of Mathematics at school. Given simple rules in an easily understood manner, a student's chances of passing an examination can be increased from a fail level to a good pass.

Need.
We have first to remind ourselves that qualifications in Maths and English are the most required subjects by employers. But they are fundamental not only to our work but to our lives as well. Our ability to calculate V.A.T., or interest on Hire Purchase, or just straight forward money calculations has to remain with us after the school exams. So, although this book is aimed at the 70% or more of students who take the lower papers in maths, IT IS ALSO AIMED AT THOSE STUDENTS WHO HAVE DECIDED MATHS IS TOO MUCH FOR THEM. By *Making Maths Easy* this book can help them to gain a maths exam certificate for presentation to future employers.

Whilst there are many books on the market aimed at *SELF-HELP,* none at present have provided the student who is written off as a failure in maths before the exam, with a method of gaining what amounts to over 20% of the marks with a single simple rule of proportion. That 20% plus another 20% to 30% gained on the first and easiest questions which most struggling students can manage, brings them within striking distance of a pass mark or grade well above that expected of them.

Approach.
The methods of dealing with certain subjects within this book have been tried and tested by the author for many years. If you fail to understand the basic logic of a problem, then the growth of knowledge is cut off until the basics of that problem are explained to you.
Despite the fact that much weight is given to assessing people's logical process, we use procedures for bypassing logic all the time in maths, and still arrive at the right answer. We don't have to know how to square

root a number like 11 because we can use a calculator. We don't need to know what a cosine is so long as we can apply it to trigonometry and obtain the right result. We even use the formula $I = \frac{PRT}{100}$ to calculate what is really a straight forward percentage calculation in simple interest.

So logic *ISN'T* absolutely essential to some calculations so long as we can remember the rules to apply.
The basic rules in maths are relatively simple. Remembering them is the hard part. Since visual memory is by far the most powerful way of retaining knowledge, this book emphasises all the rules. Students should always, in their written work, emphasise the key headings and rules for unlocking the information. Don't be afraid to mark your own emphasis in this book. It is a tool to be used to gain high marks in exams. It is a means to an end.

Along with other topics likely to gain good marks easily in an exam, this book will, in **Chapter 3,** introduce you to a simple to learn RULE OF PROPORTION, (which I call ***THE RULE OF X***). It will help you answer several seemingly unrelated questions using just that one rule.

The book starts off with some basic maths, since it is upon the basic maths that we build a knowledge of each successive topic. For instance, we need to know what a factor is before we can progress onto some of the basic algebra. We need to be able to divide by 10 and by 100 and so on in order to study properly topics like percentages. We need to know how to give answers in later questions to 2 decimal places or to three significant figures.
BUT MOST IMPORTANT OF ALL, these simpler basic subjects also form a part of each exam paper, and on their own, in the first few questions, they often amount to 10% of the marks.
Each topic has several worked examples for the student to refer to. Then exercises are set for them to work out using the methods shown. The answers are listed in the final chapter.

I most strongly recommend that students write off for specimen papers from the examination board. They are an excellent guide to the standard of the questions likely to be set in your exam. (see p.124).

All the topics vary in their ability to be mastered. If you master all the topics in this book you should get a grade C in the GCSE exam. AIM HIGH and you will achieve a better mark than if you aim at a lower grade.

Contents

1. The Basics

(1) MULTIPLYING AND DIVIDING with 10's, 100's, 1000's, etc.

Multiplying by 10.	$17 \times 10 = 170$	Add a zero.
Multiplying by 100.	$24 \times 100 = 2400$	Add two zeroes.
Multiplying by 10.	$1.7 \times 10 = 17$	Move the decimal point one place to the right.
Multiplying by 100.	$0.987 \times 100 = 98.7$	Move the decimal point two places to right.
Dividing by 10	$170 \div 10 = 17$	Knock off zero.
Dividing by 100.	$1700 \div 100 = 17$	Knock off two zeroes.
Dividing by 10.	$1234 \div 10 = 123.4$	Move the decimal point one place to left.
Dividing by 1000.	$143 \div 1000 = 0.143$	Move the decimal point 3 places to left.
	$1.43 \div 1000 = 0.00143$	We add in zeroes where necessary.

RULE: **WHEN MULTIPLYING, THE POINT MOVES RIGHT**
WHEN DIVIDING, THE POINT MOVES LEFT.

*Watch out for decimals like this:

Example. Multiply 26 by 0.1

Here we multiply by 1 and have 1 place of decimal in the answer since there is 1 decimal place in the sum.
So we get 2.6.

Example. Multiply 35 by 0.01

Here we do the same but have 2 decimal places in the sum.
So there are 2 places in the answer.
So we get 0.35.

Example. Divide 56 by 0.01.

The simplest way to do this is to present it like this:

$\frac{56}{0.01}$ then move the point to the right on top and bottom till 0.01 becomes 1.

Thus we get $\frac{5600}{1}$ which is 5600.

EXERCISE A1.

(1) Divide 12.7 by 100.
(2) Divide 3,569 by 100.
(3) Divide 7 by 100.
(4) Divide 17 by 1000.
(5) Divide 0.4 by 100.
(6) Multiply 0.7 by 100.
(7) Multiply 16 by 100.
(8) Multiply 100 by 0.1.
(9) Multiply 79 by 0.01.
(10) Divide 84 by 0.01.

(2) PRIME NUMBERS.

A prime number is a number that can only be divided by 1 and by itself.

So 1, 2, 3, 5, 7, 11, 13, 17, 19, 23, 29, etc. are prime numbers.

12 is not a prime number because it can be divided by 2 and by 3.

EXERCISE A2.

(1) Which of these numbers are prime numbers? — 11, 15, 19, 31, 41, 81.
(2) Name the prime numbers between 30 and 50.

(3) FACTORS.

Factors are numbers which, when multiplied together make another number.

So 2 and 4 are factors of 8 (i.e. $2 \times 4 = 8$).
and 3 and 5 are factors of 15.

In Algebra, a and b are factors of ab (ab means $a \times b$).
We will see the importance of this later.

EXERCISE A3.

(1) Name three factors of 30.
(2) What are the two factors of 3a?

(4) PRIME FACTORS.

Prime factors are factors that are also prime numbers.

So 2 and 3 are prime factors of 6.
and 2 and 3 are also the prime factors of 12 since $2 \times 2 \times 3 = 12$.

EXERCISE A4.

(1) What are the prime factors of 28?
(2) What are the prime factors of 56?

(5) PRODUCT.

A product is what we get when we multiply two or more numbers together.
So, 18 is the product of 2×9.
18 is also the product of $2 \times 3 \times 3$.

EXERCISE A5.

(1) What is the product of 3×34?
(2) What is the product of $5 \times 4 \times 3.5$?
(3) Express 24 as a product of its prime factors.

(6) SQUARE OF A NUMBER.

A square of a number is formed when that number is multiplied by itself.
So 2×2 is the square of 2. (i.e. 4 is the square of 2).
and 6×6 is the square of 6. (i.e. 36 is the square of 6).

2×2 is expressed as 2^2.
9×9 is 9^2, etc.
THE SMALL NUMBERS ARE INDEX NUMBERS or POWERS.

EXERCISE A6.

(1) What is the square of 7?
(2) What is the square of 15?
(3) How do we express the square of 15?

(7) SQUARE ROOTS

A square root is a number, which when squared gives the number you are square rooting. It is the reverse of a square.

So just as $2 \times 2 = 4$, then the square root of 4 (i.e. $\sqrt{4}$) is 2.

$\sqrt{9} = 3$, $\sqrt{49} = 7$, $\sqrt{64} = 8$, $\sqrt{9 \times 9} = 9$, $\sqrt{9^2} = 9$, $\sqrt{x^2} = x$,

$\sqrt{2 \times 2 \times 2 \times 2} = 2 \times 2 = 2^2$

EXERCISE A7.

(1) Use your calculator to find the square root of 361.

Note: enter 361 then press $\boxed{\sqrt{x}}$

(2) Use your calculator to find the square of 13.

Note: enter 13 and then press $\boxed{x^2}$.

(3) Square root 780.

(4) Square root 655.36.

(8) PERFECT SQUARES

Example. Express 1296 as a product of its prime factors. Then express it in index form.

First we divide 1296 by the lowest prime number that will go into it. Since it ends in an even number, we know it is divisible by 2. So 2 is not only a prime number, but a factor of 1296. Keep doing this until it is no longer divisible by 2, then divide it by the next lowest prime number that will go into it, and so on.

2	1296
2	648
2	324
2	162
3	81
3	27
3	9
3	3
	1

So $2 \times 2 \times 2 \times 2 \times 3 \times 3 \times 3 \times 3 = 1296$ expressed as a product of its prime factors.

and so $2^4 \times 3^4 = 1296$ expressed as a product of its prime factors in *index form.*

This gives us the ability to square root 1296 as $\sqrt{2\times2\times2\times2} = 2\times2$ and $\sqrt{3\times3\times3\times3} = 3\times3$.

So $\sqrt{1296} = 2\times2 \times 3\times3 = 36$.

***RULE:* TO SQUARE ROOT A NUMBER using this method of factorising (i.e. breaking down into prime factors), THERE MUST BE AN EVEN NUMBER OF EACH GROUP OF PRIME FACTORS.**

Here we had four 2's and four 3's.
Sometimes we are asked: What number must we multiply say 648 by in order to make it a perfect square.
Breaking it down into a product of its prime factors, it is

$$2\times2\times2 \times 3\times3\times3\times3.$$

We can halve the 3's but not the 2's, so as it stands, 648 is not a perfect square (i.e. it cannot be square rooted without a fraction). To make it a perfect square, we must have another 2 in its line of prime factors to make it an even number of 2's:

$$2\times2\times2 \times 2 \times 3\times3\times3\times3.$$

So we multiply 648 by 2 to make it into a perfect square.

There are rules for seeing if a number is divisible by various prime numbers. The ones to remember are:

***RULE:* A NUMBER IS DIVISIBLE BY 2 IF IT ENDS IN AN EVEN NUMBER.**
***RULE:* A NUMBER IS DIVISIBLE BY 5 IF IT ENDS IN 5 OR 0.**

Any other rules for 3 and 11 are more complicated than actually trying the division by 3 or 11 in the first place, and there is no rule for seeing if 7 will divide into a number other than trying it.

EXERCISE A8.

(1) Express 1008 as a product of its prime numbers.
(2) Express the above in index form.
(3) What is the least number we must multiply 1008 by to make it a perfect square?

2. Work with Fractions

(1) MIXED NUMBERS.

Where we have a number and a fraction, it is called a mixed number.

$$3\frac{1}{2}$$

(2) IMPROPER FRACTIONS.

Improper fractions are mixed numbers converted to a fraction.

$$3\frac{1}{2} \text{ becomes } \frac{7}{2}$$

***RULE:* MULTIPLY THE NUMBER BY THE BOTTOM LINE, ADD THE TOP LINE TO IT, THEN PUT THE NUMBER YOU GET OVER THE BOTTOM LINE.**

(3) ADDING AND SUBTRACTING MIXED NUMBERS

***RULE:* CONVERT THE MIXED NUMBERS TO IMPROPER FRACTIONS THEN ADD OR SUBTRACT THEM.**

So in $3\frac{1}{2} + 6\frac{3}{8}$

This $\frac{7}{2} + \frac{51}{8}$

Now we need a COMMON DENOMINATOR (a number into which 2 and 8 will go)
COMMON DENOMINATOR is 8

$$\frac{7}{2} + \frac{51}{8}$$

$\frac{28 + 51}{8}$ 2 into 8 goes 4, 4 × 7 is 28
8 into 8 goes 1, 1 × 51 is 51

$$= \frac{79}{8} = 9\frac{7}{8}$$

And $3\frac{1}{2} - 2\frac{3}{4} = \frac{7}{2} - \frac{11}{4}$

$$= \frac{14 - 11}{4} = \frac{3}{4}$$

Note: It isn't good policy to add and subtract the whole numbers first, as in the example above, problems would occur with taking away the fractions once we have taken 2 away from 3.

(4) CANCELLING FRACTIONS.

Cancelling fractions is a procedure for reducing the fraction to its smallest size. All you have to remember is that whatever you divide the top line by, you must divide the bottom line by as well. When the point is reached where no more factors can be found for both top AND bottom lines then the fraction is complete.

So $$\frac{2\cancel{0}\cancel{0}}{110\cancel{0}\cancel{0}} = \frac{2}{110} = \frac{1}{55}$$

Here we divided top and bottom by 100, and then by 2.
0's cancel easily by simply crossing one off the bottom line for each one crossed off the top line.

(5) MULTIPLYING FRACTIONS.

First cancel anything on the top line OF EITHER FRACTION with anything on the bottom OF EITHER FRACTION. You can cancel the bottom of one fraction against the top of another.
Then multiply out the top lines.
Then multiply out the bottom lines.
Then convert to mixed numbers if the top is bigger than the bottom.

So $$\frac{3}{4} \times \frac{2}{5} \times \frac{10}{21} \text{ is } \frac{\cancel{3}_1}{\cancel{4}_2} \times \frac{\cancel{2}_1}{\cancel{5}_1} \times \frac{\cancel{10}_2}{\cancel{21}_7} = \frac{2}{14} = \frac{1}{7}$$

And $$\frac{3}{4} \times \frac{7}{3} = \frac{7}{4} = 1\tfrac{3}{4}$$

The 3's cancel each other.

(6) DIVIDING FRACTIONS.

RULE: **TURN THE SECOND FRACTION UPSIDE DOWN AND MULTIPLY.**

So $$\tfrac{3}{4} \div \tfrac{1}{2} = \tfrac{3}{4} \times \tfrac{2}{1} = \tfrac{3}{2} = 1\tfrac{1}{2}.$$

(7) CONVERTING FRACTIONS TO DECIMALS.

***RULE:* DIVIDE THE TOP LINE BY THE BOTTOM LINE.**

Add as many 0's after the decimal point as is necessary.

Hence, $\frac{1}{4}$ becomes $4\overline{)1.00} = 0.25$
0.25

$\frac{1}{2}$ is $2\overline{)1.0} = 0.5$
0.5

$\frac{3}{4}$ is $4\overline{)3.00} = 0.75$
0.75

$\frac{1}{3}$ is $3\overline{)1.00000} = 0.33333.. = 0.\dot{3}$ (0.3 *recurring*)
0.33333

$\frac{2}{3}$ is $3\overline{)2.00000} = 0.66666.. = 0.6$
0.66666

MEMORIZE the 5 examples above. Not only are they part of any maths syllabus, they crop up every day in the world outside the classroom.

EXERCISE B1.

(1) Convert $1\frac{3}{4} + 2\frac{5}{8}$ to improper fractions and add them together. Give the answer in mixed number form.
(2) Subtract $3\frac{7}{8}$ from $4\frac{1}{4}$.
(3) Multiply $\frac{3}{5} \times \frac{5}{14} \times \frac{7}{12}$.
(4) Divide $\frac{7}{8}$ by $\frac{1}{2}$.
(5) What is $\frac{3}{5} \times \frac{6}{8} \times \frac{7}{9} \div \frac{1}{12}$.

(8) CONVERTING DECIMALS TO FRACTIONS.

***RULE:* PUT THE NUMBER OVER 1. THEN MOVE THE POINT TO THE END OF THE NUMBER. THEN ADD A 0 TO THE BOTTOM LINE FOR EVERY PLACE OF DECIMAL YOU HAVE MOVED THE POINT.**

After this, you can cancel the fraction to its lowest.

So 0.25 becomes $\frac{0.25}{1}$ which becomes $\frac{25}{100}$ which is $\frac{1}{4}$

And 0.125 is $\frac{0.125}{1}$ is $\frac{125}{1000}$ then cancel to $\frac{1}{8}$

(9) DECIMAL PLACES.

This indicates the number of figures after the decimal point.

So 0.567 is three decimal places
and 6.2 is one decimal place.

(10) ROUNDING UP AND ROUNDING DOWN.

This can be done with decimals OR whole numbers.

***RULE:* ROUND *UP* IF THE NEXT DIGIT OF THE NUMBER IS 5 OR MORE.
ROUND *DOWN* IF THE NEXT DIGIT IS LESS THAN 5.**

So writing 0.3486 to three decimal places it is 0.349 since the fourth (next) digit is 6.

And writing 0.3486 to one decimal place is 0.3.

(11) SIGNIFICANT FIGURES.

We use rounding up and rounding down. 0's are used after the significant number of figures required, otherwise we will lose a huge part of the number.

Writing 3494 to 2 significant figures is 3500
and 3494 to 3 significant figures it is 3490.

(12) ESTIMATION.

An estimate is a result of a calculation which has been made with numbers that are approximately those given in the question.

NOTE. Estimation must only be used where the question asks for an estimate.

So 3.99×6.02 can be written as 4×6 = approximately 24
and $\sqrt{8.98} \times (2.01)^2$ is 3×4 = approximately 12.

(13) SQUARING AND SQUARE ROOTING FRACTIONS.

RULE: **TO SQUARE A FRACTION, SQUARE THE TOP LINE AND SQUARE THE BOTTOM LINE.**
TO SQUARE ROOT A FRACTION, SQUARE ROOT THE TOP LINE AND SQUARE ROOT THE BOTTOM LINE.

So $\left(\frac{2}{3}\right)^2$ is $\frac{2^2}{3^2} = \frac{4}{9}$

And $\sqrt{\frac{4}{9}}$ is $\frac{\sqrt{4}}{\sqrt{9}} = \frac{2}{3}$

(14) THE NUMBER LINE.

This is a technique for measuring the TOTAL distance between two points which happen to be at the opposite side of 0.
It is used in (a) temperature calculations,
(b) heights above and depths below sea level,
and also (c) clock calculations.

So from $-12°C$ to $20°C$ the temperature rises from $-12°$ to 0 and from 0 to 20°, a total of 32°C.

From 30 feet below sea level to 65 feet above sea level we rise a total of 95 feet. (30 to 0 plus 0 to 65).

From 21.48 hours to 06.41 the next day we add the time before midnight (0.00 hours) to the hours after midnight, which is 2 hours 12 minutes plus 6 hours 41 minutes = 8 hours 53 minutes.

ALSO: From 9.36 a.m. to 16.11 p.m. we do the same. We add the hours up to mid-day (12.00 hours) to the hours after 12.00.
So 2 hours 24 minutes plus 4 hours 11 = 6 hours and 35 minutes.

EXERCISE B2.

(1) Express $\frac{4}{5}$ as a decimal without using your calculator.

Now do it with your calculator. 4 ÷ 5 =

(2) What is 0.76 expressed as a fraction?

(3) Express 0.4567 to three decimal places.

(4) Express 34.564 to two decimal places.

(5) Express 34.564 to three significant figures.

(6) Express 34.564 to 4 significant figures.

(7) Express 3453 to 3 significant figures.

(8) Express 3458 to 3 significant figures.

(9) Express 14.5 to the nearest whole number.

(10) Express 18.245 to the nearest whole number.

(11) Estimate $5.01^2 + \sqrt{80.95}$.

(12) Calculate $\sqrt{\frac{9}{49}}$

(13) A pearl diver dives from a cliff 30 metres above the surface of the water and swims to an oyster on the sand 8 metres below the surface. How far down has he travelled?

(14) A train leaves London at 22. 27 and arrives at its destination at 01.14. How long did the journey take?

(15) By a twelve hour clock I set off from work at 11.24 a.m. and arrive home at 1.13 p.m. How long did my journey take me?

3. The Rule of X

You may be saying you have never heard of it. You won't have! It doesn't appear in any textbooks nor is it taught in any schools as *The Rule of X.*

What is it? It is a RULE OF PROPORTION.

Throughout your maths lessons, and throughout the maths examination papers there are many subjects that come under the heading of proportion, but many are dealt with in different ways.

Here we are going to learn, in a matter of minutes, a simple rule that will help you answer all the questions based on proportion listed below. The list will include a heading and a simple example. Later in the book, as we come to study those subjects more closely, we will refer to *THE RULE OF X* which you should learn to use whenever possible.

The importance of THE RULE OF X can be seen when you look at exam papers and realise that there are questions often amounting to more than 20% of the total marks, which can be solved using this simple rule.

So what is it? Well first I'll give a simple problem and then explain what it is and how to use *THE RULE OF X* to solve it. The rule is always the same so remembering it should cause little difficulty.

Remember that in algebra X is always an unknown quantity

Question. If I pay £11.47 for 6.2 gallons of petrol, how much is it going to cost for 7 gallons.

Answer. Logically you can say that 1 gallon costs $\frac{11.47}{6.2}$

$$= £1.85 \text{ per gallon,}$$
$$\text{and 7 gallons is } 7 \times £1.85 = £12.95$$

THE RULE OF X would do it this way:

6.2 gal	=	£11.47
7 gal	=	£X

put gallons in one column, £'s in other

*Then put the corner opposite the X on the BOTTOM line and the other two corners on top. **ALWAYS!**

so we get $\frac{7 \times 11.47}{6.2} = £12.95$ for 7 gallons.

So why not work things out the logical way? The truth is, there are lots of people who can. But if you have difficulty with logic, then isn't it more logical to use a rule which you can memorize and use for lots of maths problems?

That was an easy example. There are many uses for THE RULE OF X, some simple some more difficult. But using it in all cases will mean:

(1) you will have less to remember,
(2) you can answer the questions with less work,
(3) you can answer them in less time, and
(4) *you'll get the answer right.*

So where can we use it?
REMEMBER, WE WILL COVER THESE TOPICS IN DETAIL AFTER THIS SECTION.

(1) Conversion and Ratios. (Page 22).

Convert 1755 pesetas to £s when 225 pesetas = £1

225ps = £1 *Put £'s in one column*
1755ps = £X *and pesetas in the other*

$$\frac{1 \times 1755}{225}$$ 1755 pesetas = £7.80

(2) Distance, Time and Speed (Page 26)

If I travel 96 miles in 2 hours and 40 minutes what is my speed in miles per hour?
Remember miles per hour are miles in 60 minutes.

96 miles = 160 minutes *Miles in one column*
X miles = 60 minutes *Minutes in the other*

$$\frac{60 \times 96}{160}$$ Speed is 36 m.p.h.

(3) Percentages. (Page 30).

What is 30 marbles expressed as a percentage of 150 marbles?

Remember that what we want a percentage of, always starts as 100%.

so 150 marbles = 100% gives $\frac{30 \times 100}{150} = 20\%$
30 marbles = X%

30 marbles is 20% of 150 marbles.

(4) Domestic Rates. (Before the poll tax but still applicable to business rates). (Page 29).

What is the rateable value of property the rates of which are £486.40 when the rate in the pound payable is £1.52?

Remember that rateable value is based on £1

£486.40 = £1.52 rate
£X = £1.00 rate

$$\frac{486.40 \times 1.00}{1.52}$$ Rateable value is £360

(5) Scale Drawings (Page 24 and Page 116).

Given a scale of 6 cm to 2.4 metres, what is 6.4 metres?

6 cm = 2.4 m
X cm = 6.4 m

$$\frac{6 \times 6.4}{2.4}$$ 6.4 m is scaled to 16 cm

(6) Pie Charts, Histograms, etc. (Page 63).

Construct a pie chart to represent the 2000 people in a village. Indicate accurately 400 eating boiled sweets, 250 eating toffees, and 1250 eating only liquorice allsorts. 100 eat no sweets.

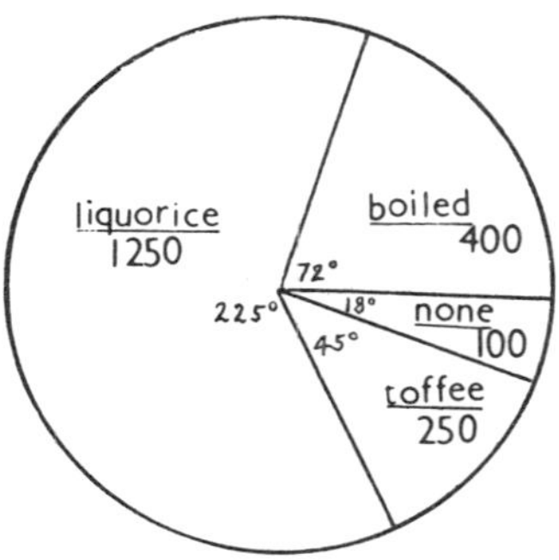

2000 people = 360°
400 people = X°

$$\frac{400 \times 360}{2000}$$ 400 people is an angle of 72°

2000 people = 360°
1250 people = X°

$$\frac{1250 \times 360}{2000}$$ 1250 is an angle of 225°

and so on. Then the angles are drawn in a circle, and the sectors are labeled with the number of people, the type of sweet, and the degrees of the angle of that sector.

(7) V.A.T. (Page 33).

If an engineer pays £3346.50 inclusive of V.A.T. at 15%, for a drill for his business, what amount of V.A.T. has he paid?

£3346.50 represents the cost of the drill + 15% Tax.

so £3346.50 = 115%

£X = 15%

$$\frac{3346.50 \times 15}{115} \qquad \text{Amount of V.A.T.} = £436.50$$

(8) Ratios of Similar Figures. (Page 34).

Given two similar triangles ABC and DEF, AB = 9 cm, BC = 8 cm, and DE is 7 cm. What is the length of EF?

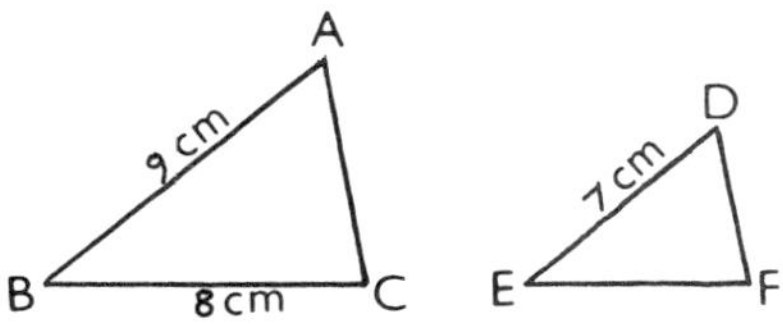

9 : 7

8 : X

$$\frac{8 \times 7}{9} \qquad \text{EF is 6.22 cm}$$

(9) Measure of Rate. (Page 25).

Example. If I apply 13.5 bags of seed on 24 acres, how many bags do I need for 58 acres?

13.5 bags = 24 acres gives $\frac{58 \times 13.5}{24}$ = 32.6 bags

X bags = 58 acres.

32.6 bags required for 58 acres.

(10) Travel and Conversion Graphs. (Page 36).

Although some other work is necessary, much in this type of question relies on your calculations of rate. E.g. rate of flow of water into a barrel, or speed of a bus covering a set distance in a given time. We will, with good reason therefore, cover this type of graph in section 12.

All these subjects will be dealt with in more detail. The purpose of this section is solely to make you realise just how many subjects are based on DIRECT PROPORTION. How to deal with them will depend on your own ability. Learn THE RULE OF X now and see if it helps in the questions you will be given.

4. Conversions and Ratios

RATIOS

Ratios split quantities into PARTS. Ratio is PROPORTION.

The sign ":" means "is to"

So using THE RULE OF X, so long as we know the *TOTAL* number of parts, we can find any part of the ratio.

Example. Find how much tin there is in 56 kg of solder that has tin and lead in the ratio 2 : 5.

7 parts = 56 kg
2 parts = X kg

$\frac{56 \times 2}{7}$ gives us 16 kg of tin.

Example. And if a solder has silver, tin, and lead in the ratio 2 : 4 : 9 what is the amount of silver in a bar weighing 3 kg?

Convert 3 kg to 3000 grams

15 parts = 3000 g
2 parts = X g

$\frac{2 \times 3000}{15}$ Silver portion is 400 g

Example. Alice, Ben and Clark have to share £20 between them in the ratio 2 : 3 : 5. How much do they get each?

There are 10 PARTS, 2 for A, 3 for B, 5 for C.

10 parts = £20
1 part = X

$\frac{1 \times 20}{10}$ 1 part is £2

MULTIPLY. A gets 2 parts = 2 × £2 = £4
B gets 3 parts = 3 × £2 = £6
C gets 5 parts = 5 × £2 = £10

Here, using THE RULE OF X makes clear that dividing the total amount to be shared by the total number of parts gives us what ONE part is worth.

Example. The angles of a triangle are in the ratio 2 : 3 : 4.
What is the size of each angle?

The angles of a triangle add up to a total of 180°
so ADD total number of parts = 9

9 parts = 180°
1 part = X°

$\frac{1 \times 180}{9}$ One part = 20°

MULTIPLY 2 parts is 2 × 20 = 40°
3 parts is 3 × 20 = 60°
4 parts is 4 × 20 = 80°

Simplifying Ratios.
A ratio of 2 : 4 can be reduced to 1 : 2 because both can be divided by a common factor which is in this case 2.
2 parts cement to 4 parts sand is the same as 1 part cement to 2 parts sand.

A ratio of 3 : 9 is simplified to 1 : 3. Both are divisible by 3.
A ratio of 10 : 24 is simplified to 5 : 12. Both are divisible by 2.

***RULE:* USE NUMBERS *YOU* KNOW WILL DIVIDE INTO THEM.**

Example. Simplify the ratio 900 : 360

We can divide by 10 and get
90 : 36
We can see 3 goes into both so divide by 3
30 : 12
Again we can divide by 3
10 : 4
And then divide by 2
5 : 2

This is as small as they will go without fractions.

Example. Simplify the ratio 90p : £2.25

First we divide by 5 pence since 5p will go into both.

This gives 18p : 45p

Now we can divide by 9

giving 2p : 5p

CONVERSION

Rates of Exchange.

Use THE RULE OF X to convert ratios of foreign currency to £s.

Example. If \$1.56 = £1 = 77.75 francs, what is \$50 in francs?

\$1.56 = 77.75 francs
\$50.00 = X francs

$$\frac{50 \times 77.75}{1.56} \qquad \$50 = 2491.99 \text{ francs}$$

Scale Models and Scale Drawings.

Use THE RULE OF X to convert scale ratios such as 1.5 cm : 12 ft.

Example. To build a model ship in the ratio 1.5 cm : 12 feet, what is the length of the model if the ship is 700 ft long?

1.5 cm : 12 ft
X cm : 700 ft

$$\frac{700 \times 1.5}{12} \qquad \text{Model is 87.5 cm long.}$$

Example. A map is drawn to the scale of 1 : 70,000. What is the actual distance represented by 2.8 cm?

1 cm : 70000 cm
2.8 cm : X cm

$$\frac{70000 \times 2.8}{1} = 196000 \text{ cm} = 1960 \text{ metres.}$$

Example. A model yacht is built 40 times smaller than the original. What length is the model yacht if the actual yacht is 136 feet long?

1 ft : 40 feet
X ft : 136 feet

$$\frac{136 \times 1}{40} = \text{model is 3.4 feet long.}$$

5. Measure of Rate

The measure of rate (not to be confused with local Rates) comes into many calculations.

Petrol is used at the rate of MILES PER GALLON.
Paint is applied at the rate of TINS PER SQUARE METRE.
Wheat seed is applied at the rate of BAGS PER ACRE.
A tank fills at the rate of LITRES PER HOUR.
Cars travel at the rate of MILES PER HOUR.
Bullets travel at the rate of FEET PER SECOND.
Wages are paid at the rate of POUNDS PER HOUR. And so on.

THE RULE OF X. We can use THE RULE OF X for any such calculation.

Example. If a car travels 48 miles on 4.6 litres of petrol, how far does it go on 20 litres at the same speed?

4.6 litres = 48 miles gives $\frac{20 \times 48}{4.6}$
20 litres = X miles

Car travels 208.7 miles.

Example. If I apply 13.5 bags of seed on 24 acres, how many bags do I need for 58 acres?

13.5 bags = 24 acres gives $\frac{58 \times 13.5}{24} = 32.6$ bags
X bags = 58 acres.

32.6 bags required for 58 acres.

Example. If a water butt holds 250 gallons of water and a tap fills it in 40 minutes, what is the rate of flow of the water?

250 gallons in 40 minutes gives $\frac{250 \times 60}{40}$
X gallons in 60 minutes

= 375 gallons per hour

Example. How many litres of paint will I need to paint the walls of a windowless store room measuring 6 m × 4 m × 2.5 m high? Two doors are 2 m × 1 m. The paint covers 3.5 m^2 per litre.

Area of walls is $6 \times 4 \times 2.5 = 60\ m^2$
Area of doors is $4\ m^2$
Total area $56\ m^2$

3.5 m^2 = 1 litre gives $\frac{56 \times 1}{3.5}$
56 m^2 = X litres

= 16 litres of paint required

6. Distance Time and Speed

There is a formula that is useful for these calculations.

$$\text{Speed} = \frac{\text{Distance}}{\text{Time}} \text{ or } S = \frac{D}{T}$$

And changing the subject, $D = T \times S$ and $T = \frac{D}{S}$

HOWEVER THE RULE OF X enables us to do these calculations *without having to remember which way round the formulas go.*

Example. A car travels 120 miles at 30 mile per hour. Then on the return journey it averages 40 m.p.h. What is the average speed for the whole journey?

*NOTE. The wrong way to do this is to add up the speeds and divide by 2. THE ANSWER WILL BE WRONG.

We must first work out the total Distance = 240 miles.
We then find the total time taken.

$$T = \frac{D}{S} \text{ so Time going } = \frac{120}{30} = 4 \text{ hours.}$$

$$\text{Time to return} = \frac{120}{40} = 3 \text{ hours}$$

Total Distance = 240 miles
Total Time = 7 hours.

$$S = \frac{D}{T} = \frac{240}{7} = 34.3 \text{ miles per hour.}$$

Remembering just **THE RULE OF X** we can still do the calculation.

Outward Journey 30 miles in 1 hour
120 miles in X hours gives $\frac{120 \times 1}{30} = 4$ hours

Return Journey 40 miles in 1 hour
120 miles in X hours gives $\frac{120 \times 1}{40} = 3$ hours

Total time for journey is 7 hours

Whole Journey 240 miles in 7 hours
X miles in 1 hour gives $\frac{240 \times 1}{7} = 34.3$ m.p.h.

Example. What is the distance of a journey when the average speed is 45 m.p.h. and the time taken is 3 hours and 20 minutes.

Remember that 45 m.p.h. is 45 miles in 60 minutes.

Using THE RULE OF X 45 miles in 60 minutes gives $\frac{45 \times 200}{60}$
X miles in 200 minutes

Distance = 150 miles

Notice we converted time to minutes to avoid a fraction of 3.33.. which would give us a somewhat inaccurate answer.

Example. What is the average speed when a journey of 140 km takes 3 hours and 30 minutes?

140 km in 3.5 hours gives $\frac{140 \times 1}{3.5}$ = 40 km.p.h.
X km in 1 hour

Example. If a rocket travels at 1200 km per hour, for 2 hours 15 minutes, how far has it travelled?

1200 km in 1 hour gives $\frac{1200 \times 2.25}{1}$ = 2700 km
X km in 2.25 hours

Rocket travels 2700 km

Example. A car travels at 70 km.p.h. for 24 minutes. Coming back it travels the same distance but takes 8 minutes longer. At what speed did it return?

First we find the distance travelled.
70 km.p.h. is 70 km in 60 minutes

70 km in 60 minutes gives $\frac{70 \times 24}{60}$ = 28 km
X km in 24 minutes

Distance travelled is 28 km.

We are given time coming back is 32 minutes.

Coming back, 28 km in 32 minutes gives $\frac{28 \times 60}{32}$ = 52.5 km.p.h.
X km in 60 minutes

Speed of return is 52.5 km.p.h.

7. Measures

RULE: Kilo means 1000 of
Centi means 100th of
Milli means 1000th of

1000 millimetres in a metre.
1000 millilitres in a litre. (1 millilitre is also 1 cm^3 i.e. 1 c.c.)
100 centimetres in a metre.
100 centilitres in a litre.
100 centigrams in a gram.
1000 metres in a kilometer.
1000 grams in a kilogram.

So to convert from one to the other we DIVIDE or MULTIPLY.

Example. What is 5,602 grams in kg? Answer 5.602 kg.
There are 1000 grams in 1 kg.
We divide by 1000 since we will have less kg than grams.

Example. What is 1.4 metres in cm? Answer 140 cm.
100 cm = 1 m.
We multiply because we want more cm than metres.

If you are unsure, and WANT TO CHECK you are not multiplying when you should be dividing use THE RULE OF X for any conversions.

100 cm = 1 m gives $\frac{100 \times 1.4}{1} = 140$ cm
X cm = 1.4 m

In this case multiply.

REMEMBER that a square 1 m × 1 m is a square metre.
that gives us a square 100 cm × 100 cm = 10,000 cm^2

And 1 m × 1 m × 1 m = 1 cubic metre = 1,000,000 cm^3

So when calculating Areas and Volumes DON'T mix measurements.

RULE: **NEVER MIX CM AND METRES WHEN WORKING ON AREAS OR VOLUMES.**

The same goes for millimetres and centimetres, feet and inches, etc.
A rectangle 30 mm × 12 cm is EITHER 3,600 mm^2
OR 36 cm^2

Convert to either mm or cm. DON'T MIX THEM.

8. Domestic and Business Rates

Rates are a local tax on householders.

RATEABLE VALUE: The house is given a Rateable Value which corresponds to its size, condition and location. For example say £300.

RATE: The rate is an annual levy which is based on the Rateable Value. At £1 Rate, the householder would pay £300 rates. That is £1 for every £1 of Rateable Value. At £1.50 Rate, the householder would pay £450. That is £1.50 for every £1 of Rateable Value.

RATES PAYABLE: The total amount payable are the actual Rates.

***RULE:* THE RATEABLE VALUE IS ALWAYS BASED ON £1.**

Example: If the rateable value of my house is £340. How much rates do I pay if the rate levied this year is £1.96 in the pound?

Rateable Value is always based on £1.00.

£340 = £1 rate gives $\frac{340 \times 1.96}{1}$

£X = £1.96 rate

Rates payable are £666.40.

Example: If my rates payable are £780.00, and the rate charged is £1.20 in the £, what is the rateable value?

£780 is what I pay at the £1.20 rate.
The question is what would I pay at the £1.00 rate.

£780 = £1.20 rate gives $\frac{780 \times 1}{1.20}$ = £650 R.V.

£X = £1.00 rate

Rateable Value = £650.

Example: If I pay £340.00 rates and the rateable value of my house is £200, what is the rate poundage charged?

£200 is what I pay at the £1.00 rate.
£340 is what I pay at the unknown rate.

£200 = £1.00 rate gives $\frac{340 \times 1}{200}$ = £1.70 Rate

£340 = £X rate

Rate of poundage is £1.70 in the £.

9. Percentages

Before covering the advantages of using the RULE OF X in percentages, we must first understand what a percentage is.

PER CENT means literally PER HUNDRED.
So 1% of anything is $\frac{1}{100}$th of it,

and 5% is $\frac{5}{100}$ths of it, and so on.

50% of an orange is $\frac{50}{100}$ths of it. (i.e. $\frac{1}{2}$ of it.)

90% of 300 marbles is $\frac{90}{100}$ths of 300 $= \frac{90}{100} \times 300 = 270$ marbles

150% of 90 is $\frac{150}{100} \times 90 = 135$

Example. My GROSS wage (that is the wage before any deductions for tax etc.) is £145 per week. What is my NET wage (that is the wage AFTER deductions) if the total deductions amount to 18% of my gross wage.

We can work out 18% of £145 and then take it away from £145.

So £145 × 18%. That is £145 × $\frac{18}{100}$ = £26.10.

and £145 − £26.10 = £118.90 NET wage.

*We should also learn to recognize that 100% − 18% is in fact 82%, and we can work out these questions in one line.

With THE RULE OF X: £145 = 100% gives $\frac{145 \times 82}{100}$
X = 82%

Which is exactly what we get if we multiply £145 by 82%

i.e. $145 \times \frac{82}{100}$ which gives us a net wage of £118.90.

***RULE:* WHATEVER YOU WANT A PERCENTAGE OF IS ALWAYS 100%**

If I want 30% of an apple then the apple is 100% before I start chopping at it. It will be 30% when I've finished chopping away the other 70% since there is 100% to start with.

Example: A bike costs £240 and I sell it for £300. Express the cost price as a percentage of the selling price.

Whatever we want a percentage of is 100%
Here we want a percentage of the selling price.
So £300 is 100%

Using THE RULE OF X, £300 = 100% gives $\frac{240 \times 100}{300} = 80\%$
£240 = X%

Percentage change.

This can either be an INCREASE or a DECREASE.

***RULE:* WHATEVER YOU ARE INCREASING OR DECREASING STARTS OFF AS 100%**

Example: Increase £220 by 5%
So if we increase my wage of £220 by 5% then we can either:

1. Work out what 5% is and add it on,

$220 \times \frac{5}{100} = £11$. So my new wage is 220 + 11 = £231

OR 2. Work out what 100% + 5% is. That is 105%

$220 \times \frac{105}{100} = £231$.

OR 3. THE RULE OF X can be used, but with straight forward percentages it is rather less useful than in many calculations.

£220 = 100% gives $\frac{220 \times 105}{100} = £231$.
£X = 105%

Example: Decrease 440 by 15%
440 is 100% so with 15% off it will be 85% of what it was.

Either: $440 \times \frac{85}{100} = 374$

Or: $440 \times \frac{15}{100} = 66$. This is 15%

and 440 − 66 = 374

Or using the RULE OF X 440 = 100% giving $\frac{440 \times 85}{100} = 374$
X = 85%

THE RULE OF X, however does make things much simpler in the questions where you are given a figure and have to calculate the percentage it represents.

Example: If I buy a car for £3500 and sell it for £4200, what is the percentage profit? (i.e. % profit on original expenditure)

Here the profit is £700 and we have to say what this is as a percentage of the COST.
So COST is 100%

$$\pounds 3500 = 100\% \quad \text{gives} \quad \frac{700 \times 100}{3500} = 20\% \text{ profit}$$
$$\pounds 700 = X\%$$

Example: I sell a video recorder that cost me £345 for £150, what is the percentage loss?

Here the loss is 345 − 150 = £195
We want this expressed as a percentage of the COST.
So COST is 100%

$$\pounds 345 = 100\% \quad \text{gives} \quad \frac{195 \times 100}{345} = 56.5\% \text{ loss}$$
$$\pounds 195 = X\%$$

Example: A car depreciates by 24% in the first year. How much is a car that costs £5520 new, worth after one year?

Here the car is worth 100% − 24% at the end of the year.
That is 76% of its original value.

$$\pounds 5520 = 100\% \quad \text{gives} \quad \frac{5520 \times 76}{100} = \pounds 4195.20$$
$$\pounds X = 76\%$$

Value of car after one year is £4195.20.

And THE RULE OF X is useful in the questions where we do not know. the amount which represents 100%

Example: After 12% of my wage has been deducted for tax I receive £198. What was my original wage?

Here we only know that £198 is 100% − 12%
so £198 is 88% of my wage.

$$\pounds 198 = 88\% \quad \text{gives} \quad \frac{198 \times 100}{88} = \pounds 225 \text{ gross wage}$$
$$\pounds X = 100\%$$

10. V.A.T.

V.A.T. stands for Value Added Tax.
It is a tax on goods or services purchased.
In Maths exams this will be simply a percentage of the value of the goods.

Example. A shop displays a sign that goods are sold 'PLUS 15% V.A.T.' What amount of V.A.T. is payable on goods costing £19.20? What is the total price of the goods?

Price of the goods is £19.20

15% of this is $19.20 \times \frac{15}{100} = £2.88$

So there will be £2.88 worth of tax to add onto the price displayed.

The total price of the goods therefore is £19.20 + £2.88 which is £22.08.

Use THE RULE OF X where the price includes V.A.T.

Example: If I buy a washing machine for £388 inclusive of V.A.T. how much tax have I paid? V.A.T. is presently 15%.

Here £388 is 100% cost of the machine + 15% V.A.T.

So £388 represents 100% + 15% which is 115% all together.

$$\begin{aligned} £388 &= 115\% \quad \text{gives } \frac{388 \times 15}{115} \\ £X &= 15\% \\ &= £50.61 \text{ tax} \end{aligned}$$

In other words we find 1% by dividing 388 by 115 and then find 15% by multiplying 1% by 15.

11. Similar Figures

Similar figures are the same shape but NOT the same size.

Similar SOLID figures are exactly the same shape

IN ALL THEIR PROPORTIONS

So a scale model of a ship is exactly the same in every detail as the real thing.

The width of the model is in the same proportion to its length as the width of the real thing is to its own length.

Similar figures have sides in direct proportion to each other.

RULE: **SIMILAR SIDES ARE IN THE SAME RATIO.**

Example. In the similar triangles below, AB = 7 cm, BC = 4 cm. DE is 9 cm. What is length of EF to 3 decimal places?

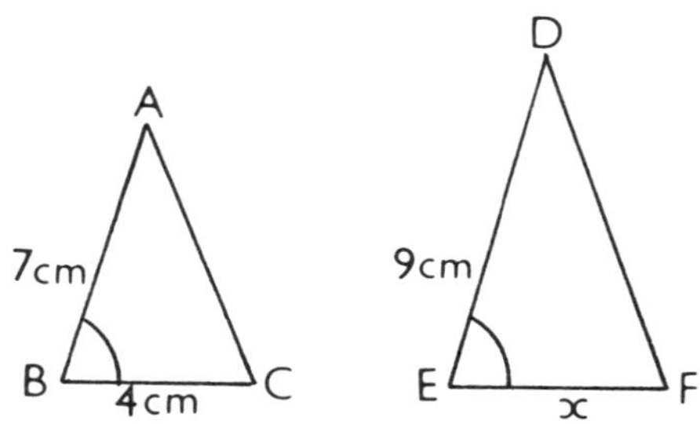

Since the sides are in the same ratio,

AB : DE
BC : EF

Small:Big

THE RULE OF X 7 : 9 gives $\frac{4 \times 9}{7} = 5.143$ cm
4 : X

EF = 5.143 cm.

Example. Two bottles are similar in all respects. The heights are 20 cm and 12 cm. The large bottle has a diameter of 7.5 cm. What is the diameter of the small one?

NOTE: Diameters are lengths of circles.

Heights are in ratio 20 : 12 = 5 : 3

Diameters are in ratio 5 : 3 They have the same proportions as the heights

THE RULE OF X $\quad 5 : 3 \quad$ gives $\frac{3 \times 7.5}{5} = 4.5$ cm.
$\quad 7.5 : X$

Diameter of small bottle is 4.5 cm.

Similar Triangles Between Parallel Lines and in Circles.

Notice that these triangles are similar. **BUT be careful** to match corresponding sides. In fig. 1, BC : CD, AC : CE and AB : DE. Here alternate angles are equal.
BUT THE TRIANGLES ARE REVERSED in fig. 2, BC : CE, AC : CD, and AB : DE. Here angles in the same segment are equal.

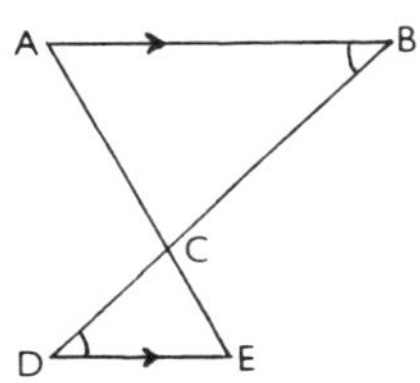

Fig. 1.

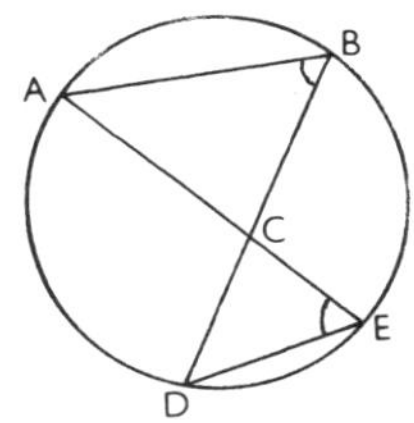

Fig. 2.

12. Time/Quantity Related Graphs

Time related graphs nearly always involve distance travelled over a period of time. However, graphs can be constructed to measure flow of liquid in an hour, or bricks laid in a day, or births in a year. The list is endless. THE RULES ARE THE SAME.

AXES.

All graphs have 2 axes for reference to each quantity.
One axis will have time, the other bricks laid.
One axis will have time, the other number of births.

One axis is horizontal, the other is vertical.

***RULE:* ALWAYS PUT THE *TIME* AXIS HORIZONTALLY.**

SCALES.

On a graph 1 square on the horizontal axis might represent 1 minute. It might represent 1 hour. One square on the vertical axis might be 10 bricks laid or 100 bricks laid.

***RULE:* WHEN PRESENTED WITH A GRAPH, WORK OUT THE SCALE ON EACH AXIS AND WRITE IT IN ALONG THAT AXIS.**

WHY? Because most of the questions can be answered quickly and simply once you have these scales worked out. Once they are written down, you don't have to keep working them out for each part of the question.

Example. The graph below represents the journey of a bus from A to B. It then travels on to C. Use the graph to find:
(1) the average speed of the bus from A to B,
(2) how long the bus waited at B,
(3) the distance of the bus from B at 9.45,
(4) the time of arrival of the bus at C,
(5) The average speed of the bus from B to C.

On the same chart, draw the graph representing the journey of a second bus that leaves C at 9.00 and travels to B at 50 m.p.h., waits at B for 18 minutes and then returns to C arriving at 12.00.
(6) At what time did the buses pass each other?
(7) How far from B were they at this time?
(8) How far apart were they at 9.36?

All the stages of working should be followed through and understood step by step.

REMEMBER that TIME related graphs can amount to 7% or more in exams. Once you have done a few graphs you should find them easy.

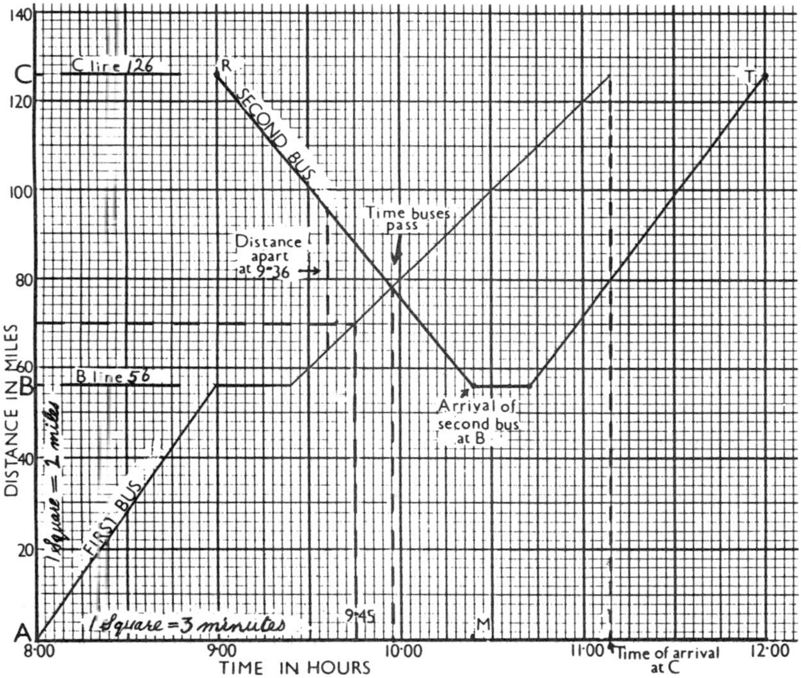

The first thing we must do is write in the scales on each axis.

1 square = 3 minutes
1 square = 2 miles.

Work out the distance of B from A and write it in. (56 miles).
Do the same for C. (126 miles).

(1) Bus sets off from A at 8.00 and arrives at B at 9.00.
Bus travels 56 miles from A to B in 1 hour

Average speed is 56 miles per hour.

(2) A horizontal graph line indicates that the bus is not travelling along the road but time is still ticking away.

Count the squares on the horizontal. There are 8

Bus is stationary for 8 × 3 = 24 minutes.

(3) The time line at 9.45 meets the graph when we draw a line vertically. A horizontal line from the point of intersection cuts the distance axis at 70 miles.
BEWARE. This is not the distance we are looking for. The bus has already travelled 56 miles to B from A. It has travelled another 14 miles since. (70 − 56 = 14). Count the 7 squares.

70 − 56 = 14 miles from B at 9.45.

(4) The bus arrived at C where the graph is in line with C. Run a line vertically down to the time axis and mark it. There are 3 squares past 11.00, so the time is 11.09.

Bus arrives at C at 11.09.

(5) To calculate the average speed from B to C we have to know the distance = 70 miles and the time taken = 105 minutes.
Read them off and make sure you understand that we read the vertical height between B and C for distance travelled and the horizontal length between 9.24 and 11.09 for the time elapsed.

Miles per hour are the same as miles per 60 minutes.

THE RULE OF X 70 miles in 105 minutes
X miles in 60 minutes $\frac{70 \times 60}{105}$

Average speed B to C = 40 miles per hour.

Drawing the second graph.

Draw a faint line or follow your ruler up the 9.00 hours line.
Draw a faint line along the C line.
Where they meet is the starting point of the graph (marked R)

Now we have to work backwards.
It is 70 miles to B from C

THE RULE OF X 50 miles in 60 minutes gives $\frac{50 \times 70}{50}$
70 miles in X minutes

The bus travels the 70 miles in 84 minutes.

Transferring this to the graph we count off 84 minutes from 9.00 to a point we mark M.
Draw a faint line up from M.
Draw a faint line across from B on the B line.
Where the lines from M and B cross indicates the TIME and PLACE of arrival at B.

Join this point to the first starting point.

The bus is stationary for 18 minutes.
That is 6 squares.
The line is horizontal because the bus isn't moving but time is passing.
Draw in the line 6 Squares long.

Then the bus travels back to C and arrives at 12.00.
Where the 12.00 line meets the C line then that point (marked T) is the end of the graph. Draw a line from the end of the horizontal waiting line to T.

(6) Where the two graphs cross the buses are in the same place at the same time. Read off the time on the time axis. The point is at 10.00 less one square, which is 9.57.

Time buses pass each other = 9.57

(7) Distance from B is read off on the distance axis. The distance we read is 78 miles. This is NOT the distance from B. That is the distance from A. So count the squares and we see there are 11 squares from B = 22 miles.

(8) The distance apart is done the same way.
Draw in a vertical line at 9.36 which crosses the two graph lines.
The number of squares on the vertical line between the graph lines is the distance apart.

Distance apart at 9.36 is 16 squares = 32 miles.

Other Graphs.

Example. Two liquids are run into an empty 100 gallon tank through 2 taps A and B. A lets liquid in at 4 gallons per minute. B lets liquid in at 5 gallons per minute. Taps A and B are turned off when the tank is full and third tap C then runs the mixture off at 8 gallons per minute. Tap A is run for 5 minutes before B is turned on.

Draw a graph to show the amount of liquid in the tank at any one time.

(1) How long does it take to fill the tank?

(2) How long has tap A been on when the tank holds 80 gallons?

(3) When is the tank half full?

(4) How long is the process from start to finish with an empty tank?

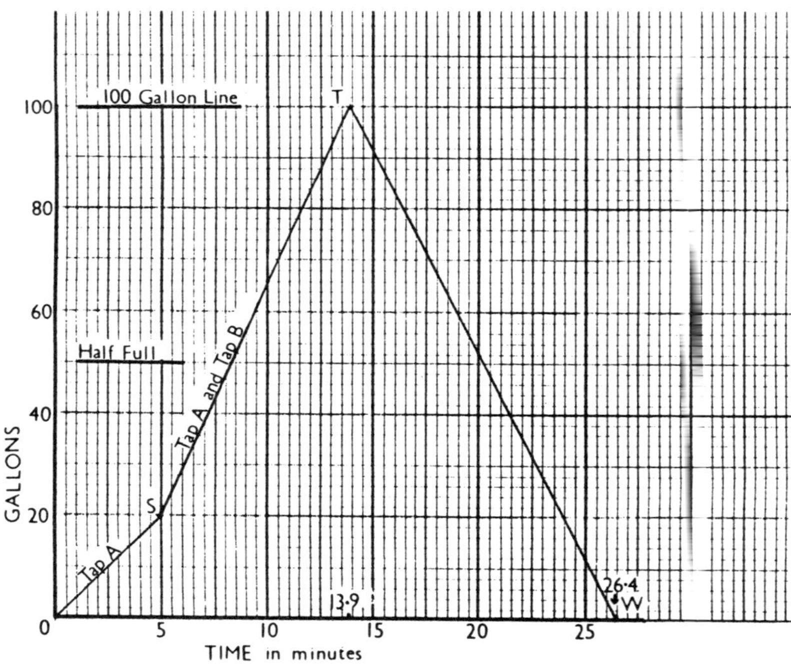

Tap A runs for 5 minutes

First work out 4 gal in 1 min. gives $4 \times 5 = 20$ gal in 5 min
X gal in 5 min.

Then mark the point S where 5 minutes crosses 20 gallons.
Draw in the first line of the graph for tap A on it's own.

Tap A and B deliver 9 gallons together in one minute.
There are 80 more gallons to run before the tank is full.

THE RULE OF X 9 gal in 1 minute gives $\frac{80 \times 1}{9} = 8.9$ min.
80 gal in X minutes

Add 8.9 minutes onto the 5 which had elapsed before B started running.
= 13.9.

(1) Tank takes 13.9 minutes to fill.
Multiply decimal part of minutes by 60 seconds 1 min = 60 sec.
= 13 minutes and 54 seconds. 0.9 min = X sec.

We can mark this point T on the graph and draw in the line for taps A and B.

Now for tap C.
THE RULE OF X 8 gal in 1 minute gives $\frac{100 \times 1}{8} = 12.5$ min.
100 gal in X minutes

So another 12.5 minutes pass on the time axis while the tank empties.
Total time taken is $13.9 + 12.5 = 26.4$ minutes.
Mark the point W, 26.4 on the time axis and draw line to meet the graph on the 100 gallon line at T.

Now we can answer the rest of the questions.

(2) Tap A has been on 10.6 minutes when the tank holds 80 gallons.

(3) Tank is half full after 8.5 minutes and again after 20 minutes.

(4) Tank is empty after 26.4 minutes.

Example. If the rate of exchange is £1 = 3.45 marks and £1 = 160 pesetas, draw a graph to convert marks to pesetas.
How many pesetas is 3 marks?
Use scale of 2 cm : 20 pesetas and 2 cm : 0.5 marks.

In this question we can see that 160 pesetas = 3.45 marks since they both equal £1.
We can therefore mark off a point 160 on the pesetas axis and draw a line up to meet the line from 3.45 marks.

So where is the other point for reference on the graph to which we can join a line?

ZERO of course! 0 pesetas = 0 marks.
Join up the 0 to the other point and we have a graph.
Read off 3 marks on the marks axis. Draw a horizontal line to intersect the graph.
Read off the answer on the Pesetas axis = 139

3 marks is 139 pesetas.

Try checking this using THE RULE OF X

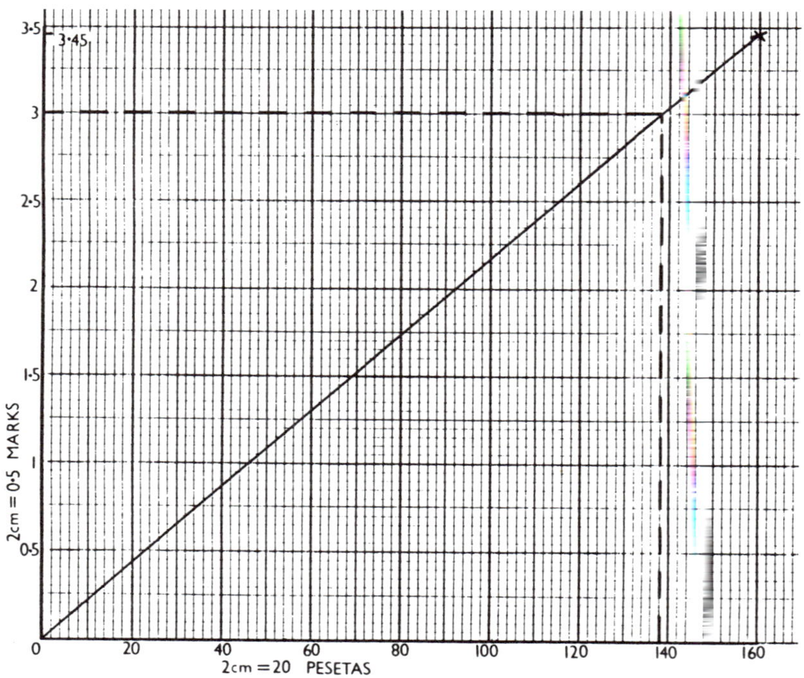

13. Exceptions to the Rule of Proportion

THE RULE OF X is used in problems of DIRECT PROPORTION. If you think the rule can be used, but the question says that something varies INVERSELY, or varies IN INVERSE PROPORTION, you must turn the X column upside down.

SUCH questions often involve energy of some kind, expended over a period of time.

*The question usually states that the quantity of one thing varies **inversely** with an increase of another. However, with practice you will be able to judge whether the problem is one of inverse proportion.

LOOK FOR A **DECREASE** IN ONE THING AS ANOTHER **INCREASES**

*Remember to *TURN YOUR X COLUMN UPSIDE DOWN**

Examples:

1. The time taken to dig a hole varies inversely with the number of men employed to dig it. If 5 men dig a hole in six days, how long will it take 3 men?

Time to dig INCREASES with DECREASE in number of men. Here, the work of three men will be less than that of five so time will not decrease in proportion, but increase in proportion. Logically, 1 man will take 30 days, so 3 men will take one third of the time = 10 days.

5 men = 6 days	invert to	5 men = X
3 men = X days		3 men = 6

so $\frac{5 \times 6}{3}$ gives us 10 days.

2. The cooking time of a joint in an oven varies *inversely* with the power used. If a cooker uses 600 watts to cook a joint in 3 hours, how long will it take to cook the joint at 800 watts?

As watts INCREASE so cooking time DECREASES. Logically, the energy used is greater, so the cooking time will be less. 1 watt will take 1800 hours, so 800 watts will take 1800 divided by 800.

With THE RULE OF X:

600 watts = 3 hours   inverts to 600 = X
800 watts = X hours   800 = 3

so $\frac{600 \times 3}{800}$. Cooking time is 2.25 hours

= 2 hours 15 minutes.

3. An emergency food pack on a life raft is sufficient to keep 6 men alive for ten days. How long will it last 8 men?

Here again, the more men there are the LESS time the food will last. There isn't a proportionate increase in food for an increase in men but a proportionate DECREASE. Therefore the food supply is inversely (or even adversely!) affected by the increase in people eating from it.

6 men = 10 days   inverts to 6 men = X
8 men = X days   8 men = 10

$\frac{6 \times 10}{8}$ Pack will last 10 men 7.5 days.

*ALWAYS REMEMBER, the **corner opposite the X** goes on the bottom line.

EXERCISE C.

(1) A woman buys a lawn mower for £130 plus V.A.T. at 15%. What is the total price she pays for the lawn mower?
(2) A Drill costs £89.70 including 15% V.A.T. What amount of tax is there in this price?
(3) A businessman can claim back from Customs and Excise the V.A.T. paid on machinery bought for his mill. If he pays a total of £9,729, how much V.A.T. can he claim back?
(4) My rates this year were £601.92 and the rate levied was £2.09. What is the rateable value of my cottage?
(5) My friend's house has a rateable value of £126.00 and he paid £204.12 rates this year. What rate was levied by his local authority?
(6) The local authority has to raise £126,000,000 next year and the rateable value of all the property in the area is £90,000,000. What rate will be necessary to raise the revenue required?

(7) What is the percentage profit if I buy a radio for £28 and sell it for £34.72?
(8) If James weighs 104 kg before his diet and weighs 81.12 kg five weeks later, what is the percentage decrease in weight? Express the new weight as a percentage of the original weight.
(9) How much was my wage if it is now £168 after a 12% increase?
(10) How much do I actually pay if I buy a car listed at £5640.00 and I get a 8% discount?
(11) Before driving to London I filled my car's petrol tank to the top. I covered 168 miles and then filled the tank to the top again. It cost me £9.80 for the petrol.
(a) If petrol is £1.72 per gallon, how many miles per gallon did I get from the car? (to 3 significant figures).
(b) How many gallons did I use? (to 2 decimal places).
(c) If 1 gallon = 4.546 litres how many litres did I use?
(d) What is the price per litre? (to 3 significant figures).
(12) A baker makes wedding cakes with measured ingredients. For a 4 kg cake he uses, along with other ingredients, 350 grams of butter and 30 grams of vegetable oil. A customer asks for a bigger cake of the same mix.
(a) How much butter will the baker need if he decides to use 75 grams of vegetable oil?
(b) What is the weight of the new cake?
(13) A car travels 90 miles in 1 hour and 48 minutes. What speed is it travelling in miles per hour?
(14) A wholesaler sells 25 kg bags of potatoes for £2.80. The retailer sells them for £3.64.
(a) What is the percentage profit made by the retailer?
(b) At what price would the retailer have to sell the potatoes to make 40% profit?
(c) If the retailer splits the bags into 5 kg lots, at what price would he mark them to make 50% profit?
(15) Certain steel has a density of 7.6 grams for 1 cubic centimetre. How many cubic centimeters are there in a girder weighing 182.4 kg?
(16) A bucket holds 6.2 litres of water. A tap fills it in 24 seconds.
(a) What is the rate of flow from the tap in litres per minute?
(b) How long will it take to fill a tank holding 46.5 litres?
(17) If $1.25 equals 2450 lire, how many lire are there in $18?
(18) If I exchange £318 to francs and £1 = 12.5 francs,
(a) How many francs do I take on holiday?
(b) How much does a present cost in pounds if I pay 54.5 francs for it?
(c) If I bring 77.5 francs back with me, how much is that in £s?

(19) A model ship is built to scale of an actual ship. The length of the funnel on the ship is 10 metres. If the model is 15.6 cm long and the length of the ship is 130 metres, what height is the funnel on the model?
(20) A map is drawn to the scale of 1 : 500,000. What is the distance between two points on the map if the distance apart is 12 km?
(21) Simplify the ratios, (a) 9 : 36, (b) 121 : 11, (c) 32 : 160, (d) 12 : 42, (e) £2.50 : 75p
(22) What is the percentage service charge in a restaurant if the original bill of £34.25 increases to £36.99 with the service charge added to it?
(23) For accounting purposes a machine costing £7,260 is presumed to depreciate at 12.5% per year. How much is it worth after 1 year?
(24) A sector of a pie chart is 54°. This represents 900 people. What angle would represent 2,100 people?
(25) (a) In fig. 1, what is the length of CE?
(b) In fig. 2, calculate the length of CE.

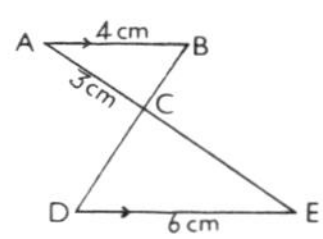

Fig. 1.

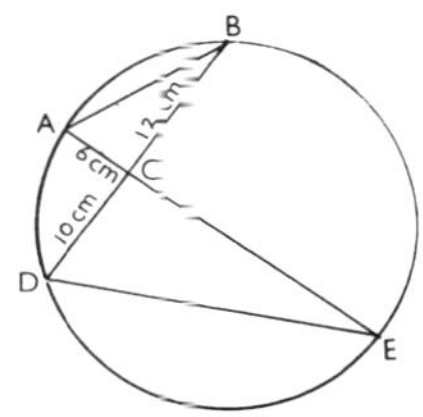

Fig. 2.

Inverse Proportion.

(26) 3 tractors can plough a field in 8 hours. How long does it take 4 tractors?
(27) 30 children eat 1 kg of toffee in 1 hour. How long does it take 12 children to eat 2 kg of toffee at the same rate?
(28) It took 12 men 56 days to build a house. How long will it take
(a) 8 men to build: 1 house, (b) 14 men to build 4 houses.
(29) If I can cut through a pole in 2 minutes, how long does it take me to cut it into ten lengths?
(30) A tap flows at the rate of 50 gallons per minute.
(a) How long will it take to fill a 80 gallon tank?
(b) If three taps are running at the same rate, how long does it take to fill a 80 gallon tank?
(31) Aunt Annie left £660 to be divided between John and Mary in the ratio 3 : 2.
(a) What are their shares?
(b) What would the shares have been if the money had included a share for Joe so that the ratio was 3 : 2 : 1?
(32) A printer has to convert a text from 54 lines per page to 40 lines per page. How many pages will he need for the conversion if he started with 94 pages?

14. Changing the Subject

In any equation we may have one or more unknown quantities. These unknowns are represented by letters of the alphabet or by the first letters of the quantity they themelves represent. For instance, in an equation about distance time and speed, we represent the distance with a D, the time with a T, and the speed with S.

The SUBJECT of an equation is that quantity that stands on its own.

So X is the subject of the equation $X = 14Y$

and I is the subject of the equation $I = \frac{PRT}{100}$

Simple Equations with ADD and SUBTRACT

An equation with one unknown quantity is a simple equation that can be solved.

Example: $3X - 12 = 2X + 4$.

***RULE:* CHANGE SIDES CHANGE SIGN.**

so in the equation: $3X - 12 = 2X + 4$ we get all the X's to one side and all the numbers to the other.

Thus: $3X - 2X = 4 + 12$ Remember, change sides change signs

so $X = 16$

Example: solve the following equation: $4X - 2 = 3X - 5$.

$4X - 2 = 3X - 5$

change sides $4X - 3X = -5 + 2$ changing signs.

so $X = -3$

Simple Equation with MULTIPLY and DIVIDE.

***RULE:* CHANGE SIDES, TOP GOES TO BOTTOM BOTTOM GOES TO TOP.**

Example: $\frac{X}{6} = \frac{2}{3}$

so $X = \frac{2 \times 6}{3} = 4$

ADD AND SUBTRACT equations with more than one unknown quantity.

Example: $X - 3Y = 7 - 3X$

Change sides $X - 3Y + 3X = 7$ Changing signs

So $4X - 3Y = 7$

Example: $3X - 4Y = 4Y + 8X - 12$

$3X - 8X - 4Y - 4Y = -12$

$-5X - 8Y = -12$

Change sides $12 = 5X + 8Y$

or $5X + 8Y = 12$ same thing

MULTIPLY AND DIVIDE equations with more than one unknown quantity.

Example: $S = \frac{VTR}{X}$

*NOTE that VTR means $V \times T \times R$ in mathematics.

Just as abc means $a \times b \times c$; and stu means $s \times t \times u$.
And 3t means $3 \times t$ and 12fgh means $12 \times f \times g \times h$.

And $3(X + 4)$ means 3 times everything in the bracket.

So in: $S = \frac{VTR}{X}$

If we have to find an equation of R in terms of S, V, T and X in other words make R the subject of the equation instead of S,

then changing sides $S = \frac{VTR}{X}$, $SX = VTR$ BOTTOM goes to TOP

And $SX = VTR$ $\frac{SX}{VT} = R$ TOP goes to BOTTOM

To find X in terms of S, V, T and R, (i.e. Make X the subject)

$$S = \frac{VTR}{X}$$

Change sides $SX = VTR$ BOTTOM goes to TOP.

and $X = \frac{VTR}{S}$ TOP goes to BOTTOM.

Mixed equations with ADD, SUBTRACT, MULTIPLY and DIVIDE

Example: $$S = \frac{VTR}{X} + 7$$

***RULE:* DON'T HAVE ANY PLUS OR MINUS SIGNS ON THE SAME SIDE AS THE SUBJECT**

To make X the Subject we must get rid of the plus on its side of the equation.

So change sides with Add and Subtract rule

$$(S - 7) = \frac{VTR}{X} \quad \text{Bracket any + or −}$$

Then change sides with Multiply and Divide rule

$$X(S - 7) = VTR \quad \text{X goes to top}$$

and $$X = \frac{VTR}{(S-7)} \quad \text{(S − 7) goes to bottom}$$

EXERCISE D.

(1) Solve the following:
 (a) $2X + 5 = 11$
 (b) $5X - 3 = 3X + 9$
 (c) $2X + 3 = 3X - 7$

(2) Simplify:
 $6a + 3b - 3a + 10b$

(3) Simplify:
 (a) $5X - 3Y = 2X + 7$
 (b) $5a - 3b = 4a - b$
 (c) $2a - 3b = 7a + 2b$

(4) Make X the subject of the equations and simplify where necessary
 (a) $3Y = X + 2$
 (b) $4Y + 2X = X + 5$
 (c) $3X - 9Y = 6$

(5) Make R the subject of the equation $I = \frac{PRT}{100}$

(6) Make f the subject of $a = ft^2$

(7) Make S the subject of the formula $9F = 27S + 2$

(8) Make P the subject of the equation $2L = \frac{3PY}{X} + 1$

(9) Make S the subject of the formula $3F = 27S + 12$ and simplify by dividing by a common factor of 3. You may refer to section 25 on factorising algebraic expressions.

(10) Make X the subject of the equation $2Y = 2X + 4$

15. Indices

*REMEMBER THIS $X + X = 2X$ and $X + X + X = 3X$

But $X \times X = X^2$ and $X \times X \times X = X$

$$a \times a \times a \times a = a^4$$
$$a \times a \times a \times a \times a = a^5 \text{ and so on}$$

Multiplying indexed numbers.

$$a^2 \times a^3 = a \times a \times a \times a \times a = a^5 \text{ (Add indices)}$$

$$\text{and } (a^4)^2 = a^4 \times a^4 = a^8$$

***RULE:* WHERE NUMBERS ARE MIXED WITH LETTERS THEN MULTIPLY OR DIVIDE EACH SEPARATELY**

Example. Simplify $5a \times 2a$.

Here we multiply 2×5 and get 10
and we multiply $a \times a$ and get a^2
so the answer is $10a^2$

Dividing Indexed numbers.

$$a^4 \div a^2 = \frac{a \times a \times a \times a}{a \times a} = a^2 \quad \text{(Subtract Indices)}$$

From this we get $a^2 \div a^4 = \dfrac{a \times a}{a \times a \times a \times a}$

$$= \frac{1}{a^2} \text{ or } a^{-2} \text{ after subtracting indices}$$

***RULE:* MINUS INDEX NUMER (say a^{-6}) MEANS $\frac{1}{a^6}$**

EXERCISE E.

(1) Simplify: (a) $2a \times 3ab$
(b) $2a^2 \times 3ab$
(c) $2a^2b^2 \times ab$
(2) Simplify: $4a^2 \div 2a$
(3) Simplify: $3a^{-4}$
(4) Simplify: ab^{-2}
(5) Simplify: $2a \times 3a^2b^2 \div 6b$

16. Formulae

Substitution

A Formula is in the form of an equation.

Something = Something Else

So $I = \frac{PRT}{100}$ and $T = \frac{D}{S}$ are both formulae.

Given the values represented by each letter we can solve these equations.

Example. Given that Y = 3a − X, find the value of X when Y = 3 and a = 12.

By substituting the values $3 = 3 \times 12 - X$

so $3 = 36 - X$

$3 - 36 = -X$

$-33 = -X$

Change sides change sign: $X = 33$

Example. If $\frac{abc}{F} = 2X$ find the value of F when a = 2, b = 3, c = 4 and X = 24.

We can make F the subject of the formula

$abc = 2XF$ so $\frac{abc}{2X} = F$

so $\frac{2 \times 3 \times 4}{2 \times 24} = \frac{1}{2}$; $F = \frac{1}{2}$

Forming an Equation

Given certain information in which we have one or more unknown quantities, we can form an equation and solve it to arrive at the value or values unknown.

***RULE:* ALWAYS USE LETTERS OF THE ALPHABET FOR THE UNKNOWN VALUES.**

Example. If 3 bananas and 4 apples cost 38p and the apples are 5p each, how much are the bananas?

Let b represent the banana.

Substitute 5 pence for each apple.

$$3b + 4 \times 5 = 38$$
$$3b = 38 - 20$$
$$3b = 18 \text{ pence}$$
$$b = \frac{18}{3} \qquad \text{1 banana costs 6 pence}$$

Other Examples of Forming an Equation.

***RULE:* IF YOU CAN'T UNDERSTAND A QUESTION WITH LETTERS IN IT THEN THINK IN NUMBERS AND APPLY YOUR ANSWER TO THE PROBLEM.**

Example. A corn merchant has W bags of wheat in his warehouse. If he sells X bags, and has 3X bags left, write down the formula for W in terms of X.

If he had 12 bags and sold 3 he would have 12 — 3 left.
If he had W   X   W − X

and we are told there are 3X left
so   W − X left = 3X bags left

$$W - X = 3X$$
$$W = 3X + X$$
$$W = 4X$$

Example. A box contains 36 apples. When X apples are sold there are Y apples left. Write down an equation for X in terms of Y.

If he sold 10 apples he would have 36 − 10 apples left
If he sold X   36 − X

and we are told there are Y apples left
so   36 − X left = Y apples left

$36 - Y = X$ or $X = 36 - Y$ same thing.

EXERCISE F.

(1) If $X = 3$ and $Y = 2$ find the value of Z in the equation $8X - Y = 2Z$

(2) If $a = 4b^2c$ what is b when $a = 9$, and $c = 4$?

(3) A boy buys 3 pens and a note book. The pens are X pence each and the notebook costs 50 pence. What is

(a) the cost of the two pens in terms of X?

(b) the total cost of the pens and the note book?

If the actual money he spends is 78 pence, how much were the pens.

17. Interest

Interest is the COST OF BORROWING MONEY. It is what we pay as a fee for borrowing money. Banks pay us interest on money we lend to them. Oh yes! They aren't just paying us for the pleasure of looking after our money. When we put money in a DEPOSIT ACCOUNT we are lending the bank money so that they in turn can lend it out to people who want to borrow some money. They charge more interest to the borrowers than they pay to the depositors, thus making a nice profit. By law they are obliged to keep one eighth of the amount they borrow from depositors on hand for covering the paying out of withdrawals.

Banks pay COMPOUND INTEREST and charge Compound Interest. Compound interest simply means that each year the interest is added onto the loan and so the next year the loan increases. Therefore the interest will be on a different amount each year. However, there are other types of Interest associated with other forms of borrowing.

*Interest is a PERCENTAGE of the loan (amount borrowed). It is usually calculated on an annual basis.

Principle is the amount borrowed.

(1) Simple Interest.

Simple Interest is calculation of a given percentage of a loan, and it is the same amount each year. The interest is not added to the loan until the loan ends, so no interest accrues on the previous year's interest. That is why it is called Simple Interest because the calculations are not complicated with the Principle increasing each year. Though there is nothing really complicated about Compound Interest either!

***RULE:* You should learn the following formula** $\mathbf{I = \frac{PRT}{100}}$

where P is the Principle, R is the Rate % of interest, and T is the number of years it is borrowed (Time).

For example: If I lend £600 for 4 years at 8% Simple Interest, then I can expect to get Interest amounting to:

$$I = \frac{600 \times 8 \times 4}{100} = £192 \text{ interest.}$$

WHY $\frac{PRT}{100}$?

Since I can expect to get 8% of £600 each year,

I will get $600 \times \frac{8}{100}$ for one year, and 4 times that for 4 years.

The formula $I = \frac{PRT}{100}$ is more use if we are asked to find the % Rate of Interest, or the Time it is invested.

Example: For what time does £110 need to be invested at 5% per annum Simple Interest, to amount to £154?

By changing the subject we get:

$$T = \frac{100 \times I}{PR}$$

The Interest is £44

so $T = \frac{100 \times 44}{110 \times 5} = 8$ years.

However, complications arise when we get a question like:

Example: What sum would amount to £403.20 after two years at 6%, Simple Interest?

Using $I = \frac{PRT}{100}$ involves a long and complicated sum.

*WE DO IT THIS WAY:

Interest for 2 years at 6% is the same as for one year at 12%, so the total interest will be 12%

£403.20 is the amount after adding interest, so it is the original amount invested plus 12%. So it is 100% + 12% (i.e. 112%).

Using THE RULE OF X

£403.20 = 112% gives $\frac{403.20 \times 100}{112} = £360$

£X = 100%

£360 was the sum originally invested.

Compound Interest.

A loan (Principle) under this scheme gets interest each year and that interest is added to the Principle so that next year the interest is calculated on a new larger sum.

Before I give an example, I will now add a Rule for you to remember which relates back to the section 1 on dividing by 100.

***RULE:* If 90% is $\frac{90}{100}$ it is $\frac{9}{10}$ which is 0.90 of something.**

$$1\% \text{ is } \frac{1}{100} = 0.01 \text{ of something.}$$

$$9\% \text{ is } \frac{9}{100} = 0.09 \text{ of something.}$$

$$15\% \text{ is } \frac{15}{100} = 0.15 \text{ of something.}$$

$$112\% \text{ is } \frac{112}{100} = 1.12 \text{ of something,} \quad \text{and so on.}$$

Example: If I invest £400 for three years at 10% compound interest what is the total amount at the end of the loan?

First year I have £400 + 10% of £400 as interest.
In other words I have 100% + 10% = **110% after 1 year**

*Now remember that 110% is $\frac{110}{100}$ ths = 1.10

So $\quad$ £400 × 110% is £400 × $\frac{110}{100}$

Which is $\quad$ £400 × 1.10 = £440

The second year I have £440 × 1.10 = £484

And the third year: £484 × 1.10 = £532.40 after 3 years.

So what can we see from this? A formula! Right!
The above example could be written:

£400 × 1.10 × 1.10 × 1.10 couldn't it?

That's multiplying by 1.10 for each year of the loan.

In other words $1.10^3 \quad R^3$

So for 5 years it would be:

$$£400 \times 1.10^5 \qquad P \times R^5$$

And for 20 years it would be:

$$£400 \times 1.10^{20} \qquad P \times R^{20}$$

***RULE:* Total amount after n years is P R^n**

where P is Principle and R is the accumulated %age
i.e. 100% + the % Rate of interest.
n is the number of years.

So for 8 years at 8% it will be PR^8 which is $P \times 1.08^8$

and for 4 years at 15% it will be PR^4 which is $P \times 1.15^4$

(3) Depreciation.

This is exactly the same as Compound Interest only instead of increasing the principle, it decreases at a given rate.

Example: A car costing £6000 depreciates at 20% per annum. How much is it worth after three years?

It will be worth 100% − 20% = 80% of its value after 1 year, which we represent as 0.80 of the value, since 80% is $\frac{80}{100}$

so £6000 × 0.80 × 0.80 × 0.80 = £3072 after 3 years.

Example: A man weighs 212 kg and on a diet he loses 4% of the previous week's weight each following week. How much does he lose in 8 weeks?

He loses 4% so he weighs 96% the next week. That is 0.96 of the previous weight.

so $212 \times 0.96^8 = 152.93$ kg weight after 8 weeks.

So he loses the difference between what he weighed at the start and what he weighed at the end of the diet.

212 − 152.93 = 59.07 kg weight loss in eight weeks.

Some calculators have a Y^x key which can be used to simplify sums to high index powers.

212 × 0.96 Y^x 8 = 152.93

(4) Hire Purchase.

Buying goods on instalments is Hire Purchase.
Buying goods on credit is a Credit Sale.

They work the same, except that with H.P. you don't own the goods until the final payment is made, and with Credit Sale, you get title to the goods immediately but borrow the money from the vendor and pay it back over a period of weeks or months.

A DEPOSIT is usually paid, and then the balance **which includes interest on that balance** is paid in instalments.
It usually costs more than paying cash and the company selling the goods has to tell you what the cash price is and what the total H.P. price is. They also have to tell you the ANNUAL PERCENTAGE RATE (A.P.R.) of interest they are charging you for giving you credit on the balance.

Example: I buy a motorbike costing £450. I pay 20% deposit and then pay the rest in 24 instalments over the next 24 months. The total price paid is £540. What interest do I pay on the credit part of the purchase and what is the annual rate of interest being charged (i.e. what is the A.P.R.?).

Cash price of the motorbike is £450.
Total price is £540.
Cost of the credit is £90.
Amount of credit is £450 less 20% of £450 paid as deposit

$= 450 - (450 \times 0.20)$
$= 450 - 90$
$= £360$ credit.

The interest on £360 credit is £90

so £360 = 100% gives $\frac{90 \times 100}{360} = 25\%$
£90 = X%

A **total** of 25% interest paid over 2 years is the equivalent of 12½% per year on the balance of £360.

EXERCISE G.

(1) For how long would £60 have to be invested at 10% simple interest to amount to more than £100? How much would the total be?
(2) What rate of simple interest would make £950 into £1149.50 in three years?
(3) What amount invested at 6% simple interest would become £496 in 4 years?
(4) What sum would amount to £960 after 4 years at 5% simple interest?
(5) For how long would £60 have to be invested at 10% compound interest to amount to more than £100? What would the total be?
(6) A bank charges 13% compound interest on a loan to a firm for 5 years.
(a) How much has to be repaid at the end of that period if the loan is £2600?
(b) If the firm pays back £800 at the end of each year, and interest is then charged on the balance, how much is still owed at the beginning of the third year and how much has to be repaid all together?
(7) If a radio active substance loses half of its radio activity every 2 years, what fraction of the original radio activity will be left after 10 years?
(8) A £700 washing machine depreciates at 15% per year. What is its value after 4 years?

18. Income Tax

Tax on earned income is devised so as to tax at a higher rate those who earn the most money. It is done in three stages.

(1) Everybody is allowed to earn some money which is not taxed. This is called TAX FREE PAY. This Tax Free Pay is an allowance made to the person earning the money, and it varies according to whether that person is SINGLE, MARRIED, or has CHILDREN.

(2) After the allowance (Tax Free Pay) has been deducted from the total earnings, the remainder is TAXABLE INCOME. This part of the income has a given percentage (say 25%) taken off it as tax.

(3) If the Taxable Income exceeds a certain amount, (say £20,000) then anything over that level is taxed at a higher percentage than the first £20,000. (Say a 40% rate of tax).

The amount of the personal allowance and the rates of tax vary from year to year according to what the government decides to allow and to charge.

Example: The tax free allowances for a married man is £4,100, and for a single man, £3,100. The lower rate of tax is 25%, the higher rate on anything over £20,000 of taxable pay is 40%. Calculate the taxable income, and the amount of tax paid each year by the following:

(a) a single man earning £212 per week,
(b) a married man earning £25,000 per year.

(a) The single man gets £212 × 52 in a year = £11,024. His tax free pay is £3,100, so if we take this from his yearly earnings we are left with £7924 taxable income at 25p in the £ (i.e. 25%).

$$\begin{array}{ll} £7924 = 100\% & \text{gives } \dfrac{25 \times 7924}{100} = £1981 \text{ tax} \\ £X \ = 25\% & \end{array}$$

[Or remember that 25% is 0.25, so £7924 × 0.25 = £1981]

(b) The married man earns £25,000 per year. From this we can deduct his Tax Free Pay allowance of £4,100.

25,000 − 4,100 = £20,900 taxable income

The first £20,000 of this is taxed at 25%

$$\begin{array}{ll} £20{,}000 = 100\% & \text{gives } \dfrac{20{,}000 \times 25}{100} = £5{,}000 \\ £X \ = 25\% & \end{array}$$

[Or remember that 25% is 0.25, so £20,000 × 0.25 = £5,000]

Anything above £20,000 is taxed at 40%

There is £900 to be taxed at 40%

$$\begin{array}{ll} £900 = 100\% & \text{gives } \dfrac{40 \times 900}{100} = £360 \\ £X \ = 40\% & \end{array}$$

[Or remembering that 40% is 0.40, £900 × 0.40 = £360]

So his total tax is £5,000 + £360 = £5,360

19. Gas and Electricity

Both Gas and Electricity are sold in units measured by a meter.

The usual way of charging the customer is to charge them a
STANDING CHARGE, and then
So much per UNIT used.

Difficulty can arise in (1) the METER READING.
(2) the COST PER UNIT

(1) The meter reading is always increasing as the units used are added onto it. So to find out how many units have been used in a quarter, the meter is read, then the previous (lower) reading is taken from that reading. This leaves the difference of units used since the last reading.

(2) The other thing to beware of is that when the cost per unit is given as say 43.5p per therm, this means £0.435, and NOT £43.50p. Divide by 100 to get pence to pounds.

Example. A house's gas bill consists of a standing charge of £10.50 plus the cost of gas used for which there is a charge of 43.5p per unit. How much is the bill when 212 units are used?

*First REMEMBER that 43.5p per unit means £0.435

£0.435 × 212	=	£92.22
Standing charge	=	£10.50
Total bill		£102.72

Example. An electricity bill consists of a standing charge of £7.25 plus the cost of electricity used at 6.20p per unit. Economy units are charged at 60% of the price of ordinary units. How much is the bill when 780 ordinary units and 425 economy units are used?

*First REMEMBER that 6.20p is NOT £6.20 but £0.062

so Economy units cost $0.062 \times \frac{60}{100}$ = £0.0372 each

so	Standing Charge	=	£7.25
then	780 ordinary units × 0.062	=	£48.36
and	425 economy units × 0.0372	=	£15.81
	Total bill		£71.42

REMEMBER. If you work out the answers in these questions without converting the price per unit from pence to pounds

YOUR ANSWER WILL BE IN PENCE.

EXERCISE H.

(1) A gas bill is made up of a standing charge of £9.80 plus the cost of gas used at 35.6p per therm. What is the gas bill for a house using 325 therms?

(2) David's bills for the year include

Weekly rent of £22 per week,

Quarterly gas bills of £87.50,

Annual rates. The rateable value of his apartment is £220 and the rate is £1.65 in the £.

(a) How much are these bills?

(b) What is his total bill for the year?

(3) A single man gets a tax free allowance of £3,450. A married man's is £4,670. The rates of tax are 24% on the first £16,700 and 40% on anything over that.

(a) What is the take home pay of a single man earning £204 per week?

(b) What is the net wage of a married man whose salary is £23,570?

(4) The gas meter reading at the beginning of a quarter was 4984 units. At the end of the quarter it was 5471 units. Gas is paid for on the basis of a standing charge of £9.80 plus 39.8p per therm used. One unit is 1.24 therms. Calculate (a) the units used, (b) the number of therms used, (c) the cost of the therms used, and (d) the total gas bill?

(5) David's hourly rate of pay is £4.40 for the first 40 hours worked during the week. Overtime is paid at one and a half times the normal rate.

(a) What is the hourly rate of pay for overtime?

(b) What is the basic weekly wage for 40 hours?

(c) What is his wage for 45 hours of work in one week?

(d) How many hours overtime has he worked to earn £255.20?

20. Averages

If we say the average height of all the people in Britain is 5.5 feet, that means there are many people below that height and many above it as well as those people who are 5 feet 6 inches tall.
The average is arrived at by:

ADDING the total height of all those people.
ADDING the total number of people in Britain.
and DIVIDING the total height by the total number of people.

This kind of average is called a MEAN.

***RULE:* ADD, ADD, DIVIDE.** Try and remember '**A**n **A**verage **D**ay'
Add what we want in the answer first.

Example. The temperatures in town for each day last week were:
22, 22, 24, 26, 25, 25, 24
What was the average (mean) temperature?

ADD temperatures = 168
ADD days = 7
DIVIDE 168 by 7 = Average of 24.

Notice the TOP line is the temperature line.

Example. If 15 kg of barley at 6 pence per kg, is mixed with 20 kg of Maize at 8 pence per kg, what price is 1 kg of the mixture?

ADD total number of pence, $(15 \times 6) + (20 \times 8) = 250$
ADD total weight = 35 kg
DIVIDE 250 by 35 = $\frac{250}{35}$ = 7.1 pence per kg.

Notice that the TOP line is the pence line. We want the answer in pence. So the top line in the calculation is the top line in the list.
So if we want an answer in pence, ADD the pence first.

Example. Marks in a test were as follows: 6 pupils got 50 marks, 5 got 55 marks, 4 got 60 and 5 got 65. What is the average mark per pupil?

What we want is an answer in marks.
so ADD total marks $(6 \times 50) + (5 \times 55) + (4 \times 60) + (5 \times 65)$
= 1140 marks all together.
ADD total pupils = 20
DIVIDE top line by bottom line $\frac{1140}{20}$ Average mark = 57.

FREQUENCY DISTRIBUTION

Lists of items or occurrences used in working out statistical data, such as averages, can be represented in charts and diagrams.

1. Frequency Table.

This marks up how one category associates with another.

Number of Plums	Weight in grams
3	50 to 59
6	60 to 69
5	70 to 79
2	80 to 89
1	90 to 99

2. Tally Chart

Same as Frequency Table except it uses tally marks.

Number of Plums	Weight in grams
𝍷𝍷𝍷	50 to 59
𝍸 𝍷	60 to 69
𝍸	70 to 79
𝍷𝍷	80 to 80
𝍷	90 to 99

3. Bar Charts

Horizontal Bar Charts.

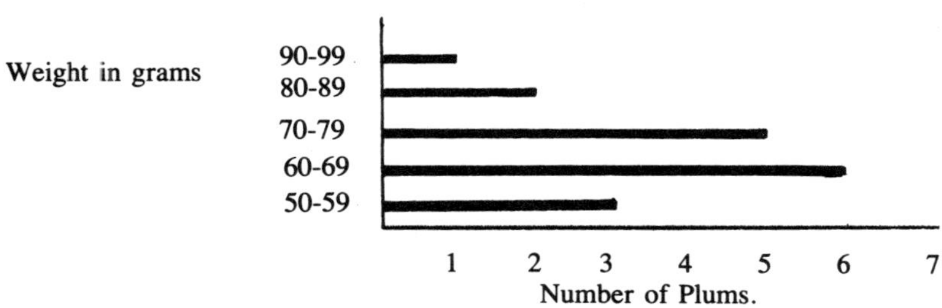

Vertical Bar Chart.

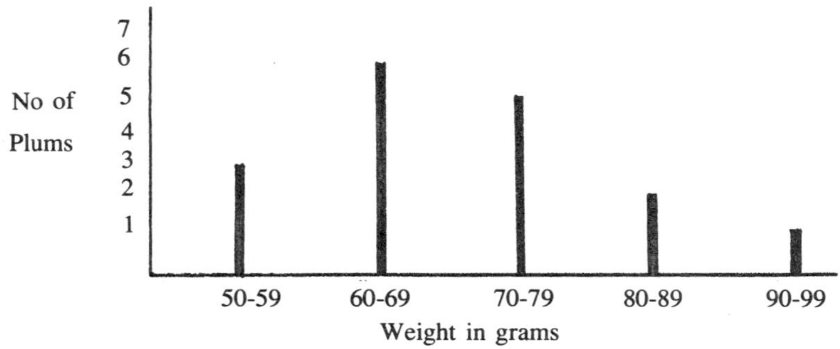

4. Histograms

These are same as vertical bar charts except the columns are touching each other.

They are sometimes called "Histograms at equal intervals" when all this means is that each column is the same width. Here the area of the column is proportionate to the number it represents. A histogram with different widths of columns will show a very distorted picture of the data it is representing.

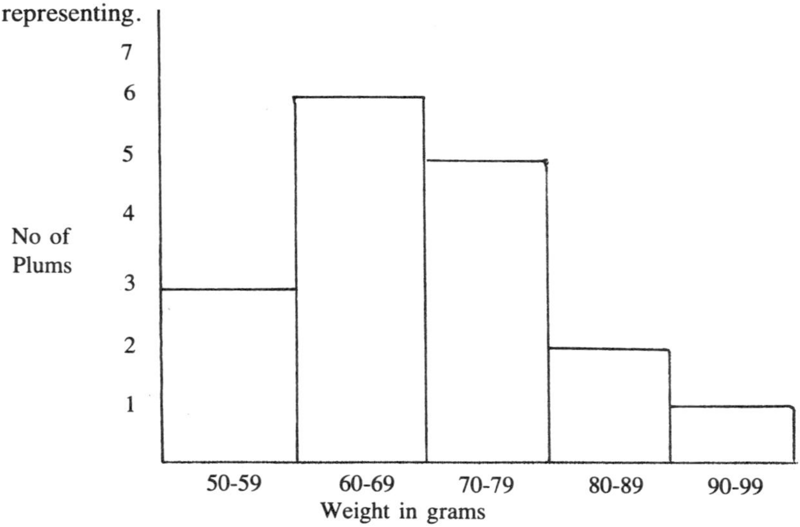

From the charts we can see that there were 17 plums weighed in our survey. Taking the average weight of each class, the total weight of the plums was (3 × 55) + (6 × 65) + (5 × 75) + (2 × 85) + (1 × 95) = 1195 grams.

The Average weight was 1195 ÷ 17 = 70.3 grams.

21. Brackets

***RULE:* ANYTHING IN BRACKETS MUST BE DONE FIRST.**

*REMEMBER: Brackets first.
then Multiplication and Division.
then Addition and Subtraction.

So in $\frac{3}{4} \times 84 - (20 - 6)$
is $\frac{3}{4} \times 84 - 14 = 63 - 14 = 49$

DID YOU NOTICE that $-(20 - 6)$ is $-20 + 6$ without a bracket?
DID YOU NOTICE that doing the multiplication before taking away 14 gives a different answer to taking 14 from 84 first and then getting ¾ of it?

***RULE:* ALL ITEMS IN A BRACKET MULTIPLIED BY SOMETHING OUTSIDE THE BRACKET MUST BE MULTIPLIED SEPARATELY.**

So $3(X + 7)$ is $3 \times X + 3 \times 7$
$= 3X + 21$

***RULE:* PLUS × PLUS is PLUS**
MINUS × MINUS is PLUS
but **MINUS × PLUS is MINUS**

If the signs are different then it is MINUS.

Example: $-3 \times -6 = 18$
$-3 \times 6 = -18$
and $-3 \times -6 \times -2 = 18 \times -2 = -36$ *DO TWO AT A TIME*

The same rule applies to division. Different signs makes minus.
so $-6 \div 3 = -2$

Examples with brackets: $3(X - 6)$ is $3X - 18$
BUT $-3(X - 6)$ is $-3X + 18$

***RULE:* FOR A BRACKET MULTIPLIED BY A BRACKET: TAKE THE FIRST NUMBER IN THE FIRST BRACKET AND MULTIPLY EVERYTHING IN THE SECOND BRACKET BY IT. THEN TAKE THE SECOND NUMBER IN THE FIRST BRACKET AND MULTIPLY EVERYTHING IN THE SECOND BRACKET BY IT.**

Example. Expand $(3X - 3)(2X + 4)$

First $\times$ First	$3X \times 2X$	is $6X^2$	Multiply numbers and then X's $2 \times 3 = 6$ and $X \times X = X^2$
First $\times$ Last	$3X \times 4$	is $+12X$	
Last $\times$ First	$-3 \times 2X$	is $-6X$	
Last $\times$ Last	-3×4	is -12	

that is $6X^2 + 12X - 6X - 12$

$= 6X^2 + 6X - 12$ Remember $-$ times $+$ is $-$

Example. Expand $(X - 6)^2$

$= (X - 6)(X - 6)$

$= X^2 - 6X - 6X + 36$

$= X^2 - 12X + 36$ This is a QUADRATIC expression

Example. Expand and simplify: $(X + 8)(X - 3)$

$= X^2 - 3X + 8X - 24$

$= X^2 + 5X - 24$

Example. Remove brackets and simplify: $4(3X + 2) - 7(X - 3)$

$12X + 8 - 7X + 21 = 5X + 29$

EXERCISE I.

(1) A student gets an average of 63% for 6 exams taken. When he has taken his 7th exam his average rises to 68%. How many marks did he get for the 7th exam?

(2) Remove brackets and simplify: (a) $4(3X + 1) - 5(X - 2)$
(b) $(X + 4)(X - 2)$
(c) $(X + 4)^2$
(d) $(X - 4)^2$

(3) Jane cycles 630 miles on her holiday. In the first five days she covered a distance of 210 miles. At the same average speed, how long does she take to cover the whole journey?

(4) Insert brackets in the following statements to make them correct.
(a) $12 \times 5 - 3 = 24$
(b) $11 \times 2 - 6 = 16$
(c) $7 + 4 \times 3 + 2 = 35$

(5) Bobby collected the following sponsorship money from 5 different people: £3, £2.50, £5.50, £4.40 and £4.60
(a) What was the average paid by each sponsor?
(b) How much did the sixth give if he averaged £4.50?

22. Fractions in Algebra

Just as we can add and subtract fractions of numbers, so we can do the same with algebraic expressions.

Example. Solve $\frac{7X - 2}{3} - \frac{X}{4} = 1$

***RULE:* PUT ALL THE NUMBERS OVER A NUMBER.**
(Those without a denominator go over 1)

***RULE:* BRACKET BINOMIALS.**
(A binomial expression has X's and numbers in it or x's and y's in it, or a's and b's, etc)

so we have $\frac{(7X - 2)}{3} - \frac{X}{4} = \frac{1}{1}$

Common denominator $\frac{4(7X - 2) - 3X = 12}{12}$

***RULE:* NOW FORGET THE DENOMINATOR.**

This is equivalent to multiplying the top line by 12. Nothing changes so long as we multiply EVERYTHING.
Then the 12's cancel.

so we are left with $28X - 8 - 3X = 12$
which is $25X = 20$

Change sides
TOP to BOTTOM $X = \frac{20}{25}$ so $X = \frac{4}{5}$

EXERCISE J.

Solve:

(a) $\frac{4X}{5} - \frac{X - 3}{2} = 3$

(b) $\frac{2X}{3} - \frac{X + 4}{4} = 4$

(c) $\frac{X}{2} - \frac{X + 2}{3} = 1$

(d) $\frac{3}{X} = 24$

(e) $\frac{24}{(X + 4)} = 8$

23. Inequality of Numbers

Inequalities are simple if they are treated as simple equations.

> means greater than
< means less than.

so $S > 10$ means S is greater than 10

Example. Express $3X - 2 > 11 + X$ in the form $X > \ldots$ Write down the least whole number value of X that satisfies this inequality.

Using the rule for equations with Add and Subtract,
CHANGE SIDES CHANGE SIGNS.

so $3X - X > 11 + 2$
$2X > 13$
$X > 6.5$

The two whole numbers at either side of 6.5 are 6 and 7
If X is bigger than 6.5, the least whole number value of X is 7 since 7 is greater than 6.5. X cannot be 6 since that is less than 6.5
so $X = 7$ satisfies this inequality.

Example. If $5X - 22 < 2X + 6$, express it in the form $X < \ldots$
What is the greatest whole number value of X which satisfies this inequality?

$5X - 2X < 22 + 6$
$3X < 28$
$X < 9.3$

The two whole numbers at either side of 9.3 are 9 and 10.

If X is less than 9.3 then it has got to be 9. It cannot be 10 since 10 is bigger than 9.3.

If the inequality is $X > \ldots$ (greater than) then we will be asked for the least whole number value so we take the next one UP.
If we are given $X < \ldots$ (less than) then we will be asked for the greatest whole number so we take the next one DOWN.

***RULE:* Greater than take the next number UP**
Less than take the next number DOWN.

Inequalities and the Number Line
Remember that if $X < -2$ then X is -3 or -4 or -5 . .
and if $X > -2$ then X is -1 or 0 or 1 . . .

24. Standard Form

Standard Form (sometimes known as Scientific Notation) is a way of writing big numbers in less space.
By using index numbers we can reduce the size of a number.

Example. $10^2 = 100$
$10^6 = 1,000,000$
$10^{18} = 1,000,000,000,000,000,000$

So 2.345×10^{18} can be written much more easily than 2,345,000,000,000,000,000

***RULE:* THERE IS ONLY EVER ONE WHOLE NUMBER BEFORE THE DECIMAL POINT.**

***RULE:* THE DECIMAL POINT MOVES THE SAME NUMBER OF PLACES AS THE VALUE OF THE INDEX NUMBER.**
Count places moved and that is the value of the index number.

Example. Express 18,000,000 in standard form

$18,000,000 = 1.8 \times 10^7$ Point has moved left 7 places

Example. Express 456,000,000 in standard form

$456,000,000 = 4.56 \times 10^8$ Point has moved left 8 places

Standard Form with Minus Index

We saw earlier that 10^{-4} is the same as $\frac{1}{10^4}$

so $6.9 \times 10^{-4} = 6.9 \times \frac{1}{10^4}$

Dividing 6.9 by 10,000 we get

$6.9 \times 10^{-4} = 0.00069$

Example. Express 0.00457 in standard form.

$0.00457 = 4.57 \times 10^{-3}$ Point has moved right 3 places

Example. Express 6.785×10^{-3} as a number.

$6.785 \times 10^{-3} = 0.006785$ Point has moved left 3 places

To change numbers to Standard Form

1. Reduce to ONE WHOLE NUMBER BEFORE THE DECIMAL POINT.
2. THE POINT MOVES AS MANY PLACES AS THE POWER.
3. If the number has got smaller by moving the point to the LEFT then the index will be POSITIVE in order to make the number bigger when it is worked back out.
4. If the point has moved RIGHT then 10 has NEGATIVE INDEX to balance an increase in the size of the number.

To change Standard Form back to numbers:

This is a simple case of multiplying by 10 as many times as its power (index number).

If the index number is negative then we put 10 and its index number underneath. This is the same as dividing by 10 as many times as its NEGATIVE index.

Add 0's when necessary for each place moved.

Example. The thickness of a sheet of paper is 1.5×10^{-4} metres
Calculate the thickness of a wad of 200 sheets.

One sheet is $\frac{1.5}{10^4}$ metres thick

which is 0.00015 m thick.

1 m = 100 cm		means we multiply 0.00015 by 100
0.00015 m = X cm		to get it to cm

so it is 0.015 cm thick.
200 sheets will be $0.015 \times 200 = 3$ cm thick.

EXERCISE K.

(1) Express 231,000,000 in standard form.
(2) Express 0.000065 in standard form.
(3) What is 4.5×10^{-4} expressed as a number?
(4) Express $11 - 3X < 54 - 10X$ in the form $X < \ldots$ What is the greatest whole number value of X which satisfies this inequality?
(5) What is the velocity of a rifle bullet that travels from the muzzle of the gun to its target 105 feet away in 8.75×10^{-2} seconds?

25. Factorising Algebraic Expressions

REMEMBER. To factorise anything we have to find COMMON factors.

Only two types of expression need be dealt with at lower level, but 2 other expressions are worth mentioning as well in (3) and (4).

(1) $4a + 10ab$

***RULE:* LOOK FOR WHAT WILL DIVIDE INTO EACH SIDE. THEN TAKE OUT WHAT IS COMMON TO EACH SIDE. BRACKET THE REST.**

$$4a + 10ab$$

is $2a(2 + 5b)$

2 and a are common factors of both numbers.

Example. Factorise $3b^2 - 12bc$

It is $3b(b - 4c)$ 3 and b are common to both.

Example. Factorise $6a^2 - 15ad$

It is $3a(2a - 5d)$ 3 and a are common to both.

(2) $4(3X + 1) - Y(3X + 1)$

***RULE:* TAKE OUT WHAT IS COMMON TO EACH SIDE. BRACKET THE REST.**
DO NOT expand the expression first

$(3X + 1)$ is common to both sides.

Since $(3X + 1)$ is in a bracket leave it in a bracket.
THEN BRACKET THE REST.

$$4(3X + 1) - Y(3X + 1)$$
$$(3X + 1)(4 - Y)$$

Example. Factorise $3a(4 + b) + 7(4 + b)$

It is $(4 + b)(3a + 7)$

Example. Factorise $9X(Y - 2) - 3(Y - 2)$

It is $(Y - 2)(9X - 3)$

Example. Factorise $2a(19 - b) + 19(19 - b)$

It is $(19 - b)(2a + 19)$

(3) $X^2 - Y^2$

This is called THE DIFFERENCE BETWEEN 2 SQUARES.

***RULE:* SQUARE ROOT EACH NUMBER THEN PUT IN BRACKETS IN THE FORM:**

$$(X + Y)(X - Y)$$

Example. Factorise $a^2 - b^2$

This becomes $(a + b)(a - b)$

Example. Factorise $X^2 - 16$

This becomes $(X + 4)(X - 4)$

Example. Factorise $a^2 - 4b^2$

This becomes $(a + 2b)(a - 2b)$

Example. Factorise $49 - c^2$

This becomes $(7 + c)(7 - c)$

Example. Factorise $4x^2 - 1$

This becomes $(2x + 1)(2x - 1)$

(4) $X^2 + 12X + 20$.

This is a quadratic expression.
A quadratic has an index power of 2.
We have seen that multiplying brackets gives us similar expressions. Now we have to get it back into brackets!

***RULE:* LIST THE FACTORS OF THE NUMBER**
PICK THE FACTORS THAT ADD UP TO THE NUMBER IN FRONT OF THE X (That number by the way is called a coefficient). THEN PUT EACH FACTOR (WITH ITS SIGN) IN A SEPARATE BRACKET WITH AN X.

In this case, $X^2 + 12X + 20$,

list the factors of 20,
then ask which ones add up to 12, and multiply to make 20

Factors of 20 are: 1×20,
2×10,
4×5

2 and 10 add up to 12. $2 \times 10 = 20$

so $(X + 2)(X + 10)$ is the solution

Example. Factorise $X^2 - 7X - 30$

Factors of 30 are 1×30,
2×15,
3×10,
5×6.

NOW $+3 - 10$ add up to -7 and $3 \times -10 = -30$

so $(X + 3)(X - 10)$ is the solution.

Example. Factorise $X^2 - 14X - 15$

The factors of 15 are 1×15,
3×5.

$-15 + 1 = -14$ and $-15 \times 1 = -15$

so $(X + 1)(X - 15)$ is the solution.

Example. Factorise $X^2 + 8X + 12$

Factors of 12 are 1×12
2×6
3×4.

and $2 + 6 = 8$ and $2 \times 6 = 12$

so $(X + 2)(X + 6)$ is the solution.

Example. Factorise $X^2 - 9X + 8$

Factors of 8 are 1×8,
2×4.

and $-1 - 8 = -9$ and $-1 \times -8 = +8$

so $(X - 1)(X - 8)$ is the solution.

26. Probability

Probability is the likelihood of an event happening.
It is measured in CHANCES

The probability of picking an ace out of a pack of cards is 4 chances out of 52. This is written as a fraction $\frac{4}{52} = \frac{1}{13}$

The chance of picking a red ball out of a bag containing 6 red, 4 blue and 10 green balls is

$$\frac{6}{20} \text{ which is } \frac{3}{10}$$

since there are 6 chances of a red ball being grabbed out of a total of 20 balls.

The probability of throwing a 6 with a dice (this should be a 'die' but everybody calls a single die a dice and so do I) is

$$\frac{1}{6}$$

The chance of picking a red ball out of a bag of ten red balls is

$$\frac{10}{10} = \frac{1}{1}$$

Obviously this is a 1 in 1 chance which is a certainty.

Now a lot can be said about probability but it can be summed up in a simple saying which I'll write as a rule.

***RULE:* In maths AND means ADD. (e.g. 2 and 4 = 6)**
BUT NOT IN PROBABILITY!

So what does it mean in probability? IT MEANS MULTIPLY

Example. A box contains 6 white rats, 10 brown rats and 4 black rats. If one can escape at a time, and all are equal in ability,
(1) What is the probability of the first to escape being brown?
(2) If the first one is brown and it isn't recaptured, what is the probability of the second rat to escape being brown?
(3) What is the probability of the first and the second rat being brown.

(1) Probability of a brown rat is $\frac{10}{20} = \frac{1}{2}$

(2) Probability of second rat being brown is $\frac{9}{19}$ since there is one less brown rat making 9, and one less in the box of 20 making 19 in the box.

(3) Probability of a brown rat **AND THEN** another brown rat is

$$\frac{10}{20} \times \frac{9}{19} = \frac{9}{38}$$

Why? Because *AND in maths means add, BUT NOT IN PROBABILITY.*

Example. A bag contains 5 red beads, 6 blue beads, 8 white beads and 1 green bead. What are the chances of picking out a blue bead then a green bead then a red bead in that order? The first and second beads are not replaced.

Chance of a blue bead is $\frac{6}{20}$

Chance that next is a green bead is $\frac{1}{19}$

Chance that next is a red bead is $\frac{5}{18}$

So what is the chance of a blue AND THEN a green AND THEN a red.

Remember *AND in maths means add, BUT NOT IN PROBABILITY.*

so we have $$\frac{6}{20} \times \frac{1}{19} \times \frac{50}{18} = \frac{5}{114}$$

***RULE:* In probability OR means ADD.**

Example in the above question, what is the probability of a red OR a green bead being picked out?

Because red and green can be grouped as one colour since it does not matter which we pick out, we can say there are

5 chances of red and 1 of green = 6 chances out of 20.

that is $$\frac{5}{20} + \frac{1}{20} = \frac{6}{20}$$

In other words OR means ADD.

BUT watch out for catch questions.

Example. A box contains 4 white, 4 blue and 6 red counters. If the first two counters removed are blue, what is the probability of the third being red?

> Here we have a simple case of the questioner really saying there are not 14 counters in the bag but 12. We know the first two were blue, so there is no chance of them being any other colour. The question did NOT ask for the chance of a blue and then a blue and then a red. It states that a bag now contains 12 counters, 4 white, 2 blue and 6 red, so the probability of a red one is:
>
> $$\frac{6}{12} = \frac{1}{2}$$

Example. If the chance of A winning a snooker match is $\frac{2}{3}$ and the chance of B winning it is $\frac{1}{4}$ what is the chance of B winning neither of two games?

***RULE:* All chances should add up to 1**

So here we have $\frac{1}{4} + \frac{2}{3} = \frac{11}{12}$

(Note, we could have said the chance of A winning OR B winning).

So what is the other $\frac{1}{12}$?

It's got to be a DRAW!

That way, the chance of A winning OR B winning OR a draw would be 1 which is certain (no chance at all) unless the match is postponed!

The chance of B winning neither is the chance of A winning OR a draw, which is

$$\frac{2}{3} + \frac{1}{12} = \frac{9}{12} = \frac{3}{4} \text{ i.e (1 minus the chance of B winning)}$$

It is worth mentioning a few points here.

Dice.

If two dice are thrown separately then the chance of two sixes in a row is chance of a six $= \frac{1}{6}$ and then the chance of a six $= \frac{1}{6}$

and *AND in maths means add, BUT NOT IN PROBABILITY*

Chance of a six AND THEN a six is $\frac{1}{6} \times \frac{1}{6} = \frac{1}{36}$

If they are thrown together the result is the same, since they will fall as if thrown separately. How one lands is not dependent on how the other lands.

Coins.

The same applies to tossing coins. Each toss is independent of the previous toss, so six tails in a row will not alter the 1 in 2 chance of a tail coming up the seventh time.

The chances of 4 heads in a row however, would be the chance of a head, AND THEN a head, AND THEN a head, AND THEN a head.

and *AND in maths means add, BUT NOT IN PROBABILITY*

$$\frac{1}{2} \times \frac{1}{2} \times \frac{1}{2} \times \frac{1}{2} = \frac{1}{16}$$

At Least.

Where a question asks for the probability of AT LEAST a certain amount of events then we must take from 1 (100% chance) the probabilities of NOT EVER getting that number of events.

Example. A coin is tossed three times. What is the chance of AT LEAST one head?

Since there is no chance of a head if we get three tails,

The chance of tail AND THEN a tail AND THEN a tail is

$$\frac{1}{2} \times \frac{1}{2} \times \frac{1}{2} = \frac{1}{8}$$

so the chance of AT LEAST ONE head is $1 - \frac{1}{8} = \frac{7}{8}$

It becomes more difficult when we have to calculate the probability if AT LEAST 2 or more occurences.

This sort of calculation is one reason for using diagrams called probability trees.

Probability Trees.

These are a graphical rather than a purely arithmetic way of working out probability.

The diagram consists of a family tree type calculation where each branch depicting a particular probability is broken down into more probabilities as more samples or options are taken.

Example. A coin is tossed three times in a row. What is the chance that a tail will be obtained on all three occasions? What is the probability of throwing at least one head? What is the chance of throwing at least two heads?

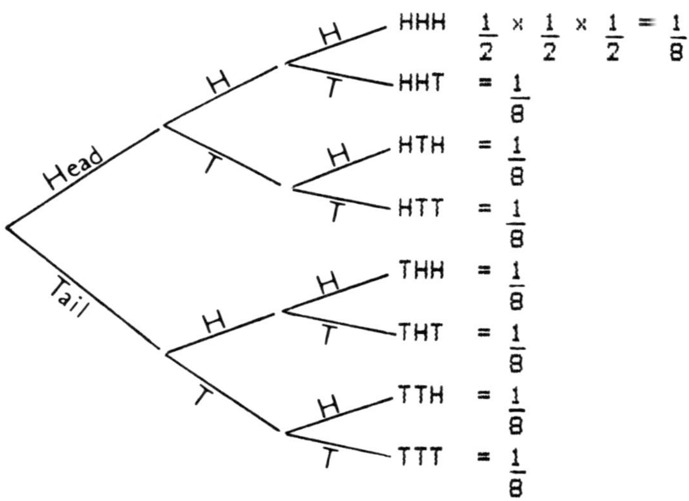

From the diagram we can see there is a ½ chance of a tail for each toss of the coin.

Chance of 3 tails. A tail AND THEN a tail AND THEN a tail is

$$\frac{1}{2} \times \frac{1}{2} \times \frac{1}{2} = \frac{1}{8}$$

The chance of at least 1 head occurs every time except the one time that 3 tails occur in the 3 tosses.

Therefore:

The chance of getting at least one head is $\frac{7}{8}$

The chance of at least 2 heads is $\frac{4}{8}$ which is $\frac{1}{2}$

Now with probability trees there are two things to remember.

(1) Multiply along the lines
(2) Add across the lines.

Example. In the above diagram, the chances of 2 heads and a tail in any order are:

$\frac{3}{8}$ Adding ACROSS the lines.

the chance of H T T in that order is $\frac{1}{2} \times \frac{1}{2} \times \frac{1}{2} = \frac{1}{8}$

following and multiplying ALONG the line.

EXERCISE L.

(1) Factorise: (a) $10X^2 - 2X$ (b) $2(a - 3b) - 4a(a - 3b)$

(2) Mrs. Jones has 3 children. What is the chance of:

(a) all three are girls?
(b) the first two are boys?
(c) only one is a girl?
(d) at least two girls?

(3) A draw for a place on a free trip is pulled out of a hat. There are 2 boys and 5 girls. The boys are John and Peter.

(a) What is the chance of a girl?
(b) What chance of it being John?
(c) What chance that the first two drawn will be girls?
(d) What chance that the first drawn is either Peter or John?

27. Basic Geometry

Angles.

Remember Angles are measured in degrees. Later we will see that degrees are split into even smaller parts called MINUTES. Minutes are split into extremely small parts called SECONDS.

1 DEGREE IS 60 MINUTES
1 MINUTE IS 60 SECONDS

Your calculator **MUST** have these. They are the DMS or °′″ key.

90° IS A SQUARE ANGLE

Acute angle Less than 90°   *Obtuse angle More than 90°*   *Reflex angle More than 180°*   *Right angle 90°*

Remember 2 right angles make a straight line.

RULE: **ANGLES ON A STRAIGHT LINE ADD UP TO 180°**

Remember 4 right angles fill a complete circle.

RULE: **ANGLES ABOUT A POINT ADD UP TO 360°**

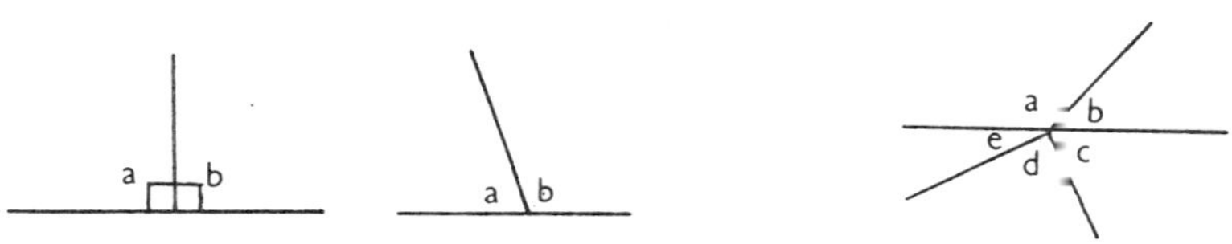

$a + b = 180°$ $a + b + c + d + e = 360°$

When 2 lines cross each other,

RULE: **OPPOSITE ANGLES ARE EQUAL.**

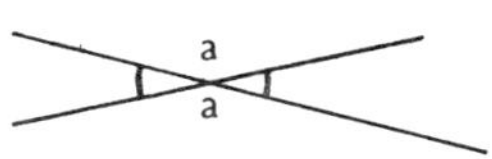

28. Properties of Triangles

Types of triangle.

(1) Acute angled. All the angles are less than 90°

(2) Obtuse angled. One angle is greater than 90°

(3) Right angled triangle. One angle is 90°
The side opposite the right angle is the HYPOTENUSE.

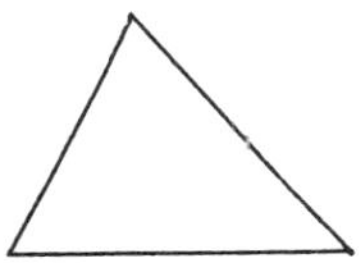

Acute Angled

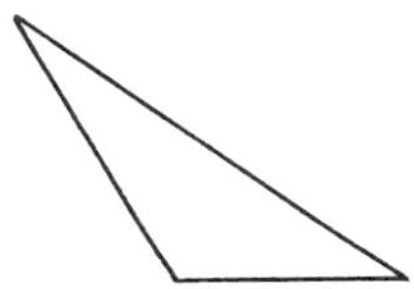

Obtuse Angled

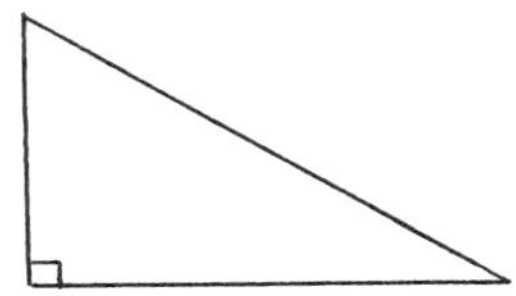

Right Angled

(4) Isosceles. Two sides are equal in length.
The base angles are equal.

(5) Equilateral. All the sides are equal in length.
All the angles are equal. They are 60°

Isosceles

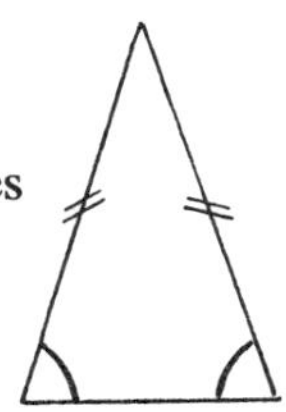

Equilateral

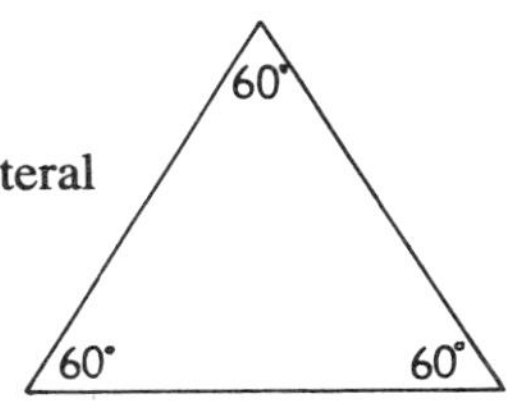

Remember the following.

(1) **ALL INTERNAL ANGLES OF A TRIANGLE ADD UP TO 180°**

(2) The **EXTERIOR** angle (marked f) is equal to the sum of the two opposite internal angles.

f = a + b
Why? Because
a + b + c = 180°
and f + c = 180°
so f = a + b

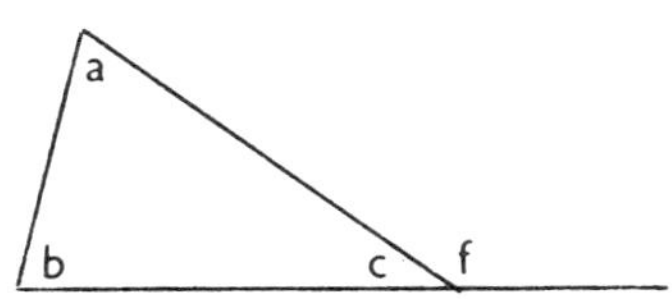

(3) The perpendicular from the apex (the top) of an isosceles triangle bisects (cuts in two equal parts) the base line.

***RULE:* A PERPENDICULAR IS AT RIGHT ANGLES TO A LINE.**

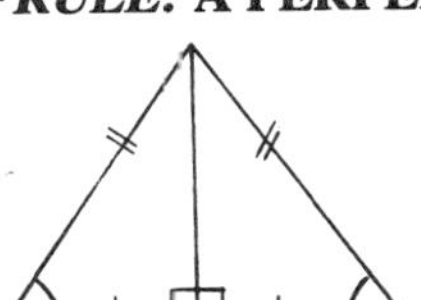

29. Parallel Lines

Parallel lines are the same vertical distance apart at any point in their lengths. They are applied to geometry when a line connects or cuts through them.

Remember the following rules.

(1) **ALTERNATE ANGLES ARE EQUAL.**

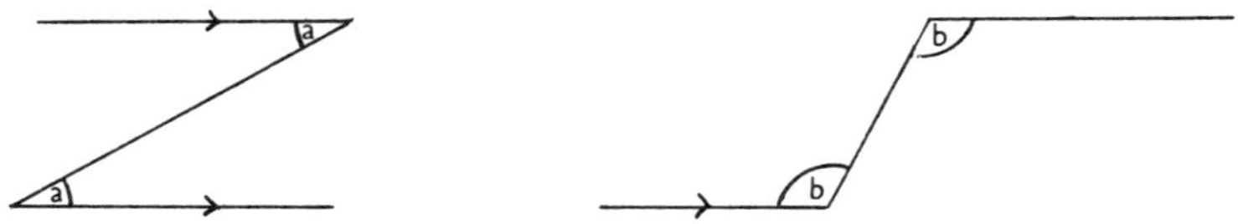

Watch for the Z shape.

(2) **CORRESPONDING ANGLES ARE EQUAL.**

(3) **OPPOSITE ANGLES ARE EQUAL.**

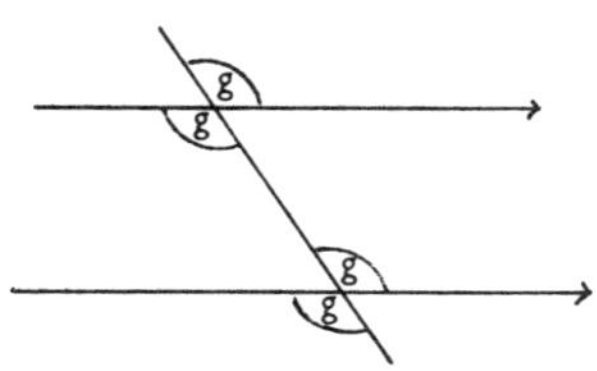

(4) **INTERNAL ANGLES ADD UP TO 180°**

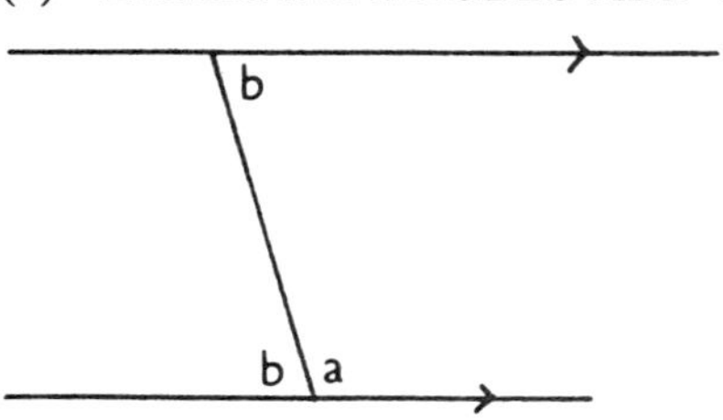

Why? Because a + b = 180° on a straight line.

b indicates alternate angles

Example. Find angles a, b, c, d and e.

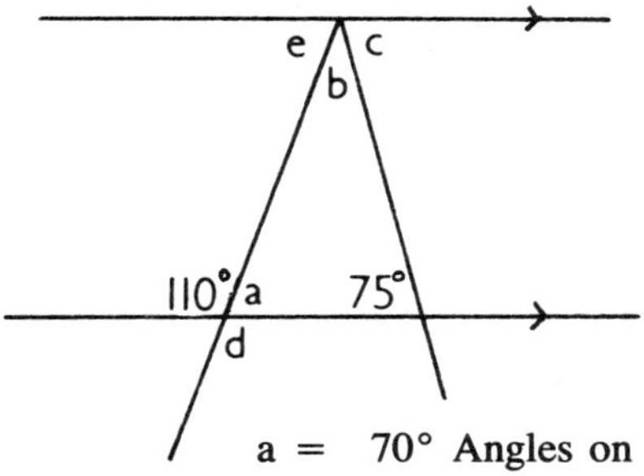

a = 70° Angles on a straight line.
b = 35° 3 angles of a triangle = 180°
c = 75° Alternate angle
d = 110° Opposite angle
e = 70° Internal angles = 180°

Don't be afraid to extend lines in order to do your calculation. You can mark angles with your own letters of the alphabet to help.

Example. Find *a* in the diagram below.

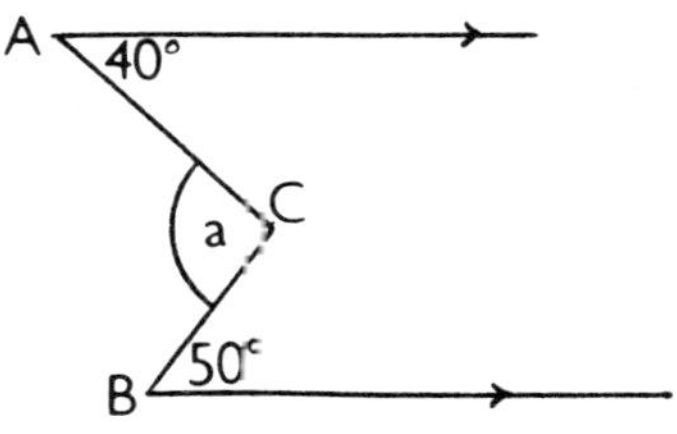

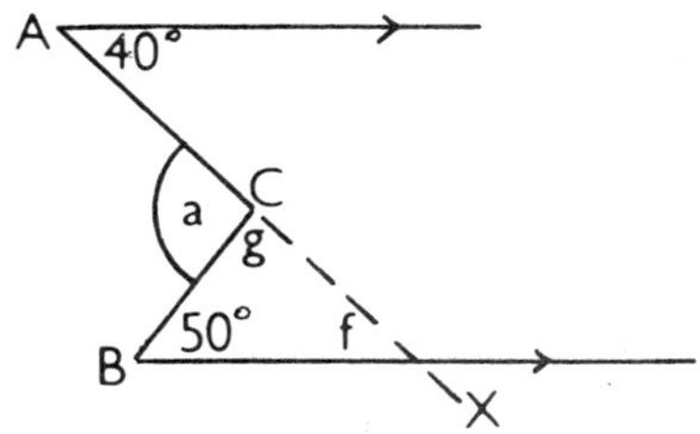

Extend AC to X
Mark angles f and g

then f = 40° Alternate angle
g = 90° Angles of a triangle = 180°
a = 90° Angles on a line = 180°

30. Pythagoras

In a right angled triangle, the HYPOTENUSE is the side opposite the right angle.
Calculations can be made to find the length of a side or the hypotenuse given two out of the three lengths.

RULE: **The theorem is: THE SQUARE ON THE HYPOTENUSE IS EQUAL TO THE SUM OF THE SQUARES ON THE OTHER TWO SIDES.**

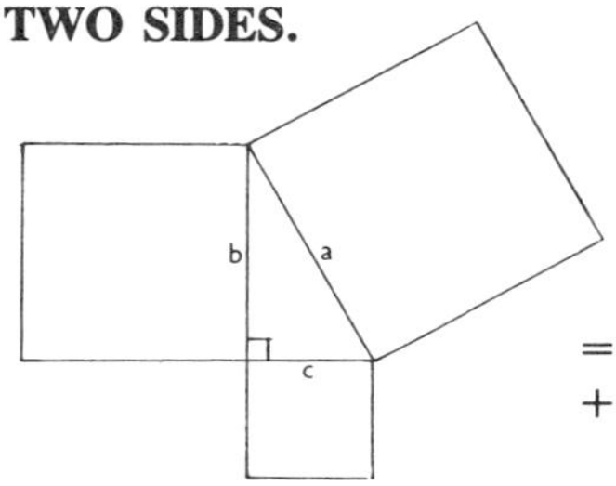

Area of square on a
= Area of square on b
+ Area of square on c

Explained, this simply means that if you square the length of the hypotenuse then the result is the same as adding together the squares of the lengths of the other two sides.

So given the diagram below,

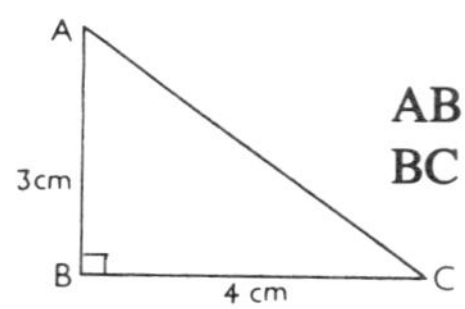

AB = 3 cm
BC = 4 cm

So Square AB and BC $\quad AB^2 = 9$

$BC^2 = 16$

Now $AC^2 = AB^2 + BC^2$ so $AC^2 = 25$

Therefore $\quad AC = 5$ cm

RULE: **RECOGNIZE A 3, 4, 5 TRIANGLE.**
It crops up in every exam!

Label the hypotenuse a, and the sides b and c so that we see imply that

$$a^2 = b^2 + c^2$$

Then from this equation we know that if we are only given two of these measurements we can find the other.

$$c^2 = a^2 - b^2$$

So in the diagram below, if a is 13 cm, b is 5 cm, what is the length of side c?

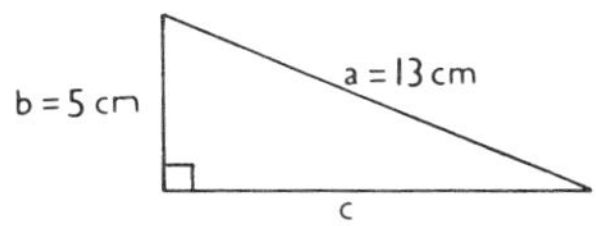

$$c^2 = a^2 - b^2$$
$$c^2 = 169 - 25$$
$$c^2 = 144$$
$$c = 12 \text{ cm.}$$

RULE:* RECOGNIZE A *5, 12, 13 TRIANGLE.

It crops up in *almost* every-exam!

Both the 3,4, 5 triangle and the 5, 12, 13 triangle can be scaled up (or down).

3,	4,	5,	5,	12,	13
6,	8,	10,	10,	24,	26
9,	12,	15		and so on	

Practise by checking that $26^2 = 24^2 + 10^2$
and that $15^2 = 12^2 + 9^2$

Not all triangles will work out as exact numbers.

Example. In the triangle below find the length of the side marked c to 3 decimal places.

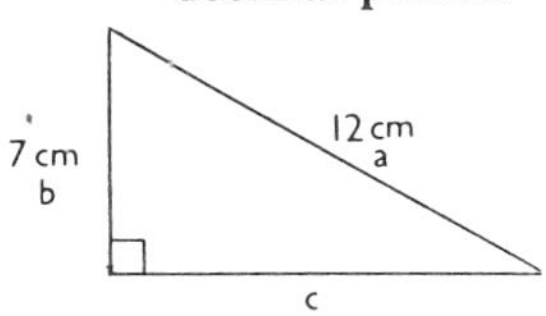

$$a^2 - b^2 = c^2$$
$$144 - 49 = c^2$$
$$95 = c^2$$
$$c = 9.747 \text{ cm}$$

Mark the hypotenuse a
Mark the other side b

EXERCISE M.

(1) In a right angled triangle ABC, AC is the hypotenuse. AC is 10 cm, and AB is 6 cm. How long is BC?
(2) In a right angled triangle whose hypotenuse is 39 cm, and one side is 15 cm, how long is the third side?
(3) Town A is 18 km due south of B, and town C is 27 km due east of A. How far apart are towns B and C to 3 decimal places?
(4) In an isosceles triangle ABC, AB = AC. The apex angle is 38°. What is the size of the base angles?
(5) What is the size of angles a, b and c in the diagram?

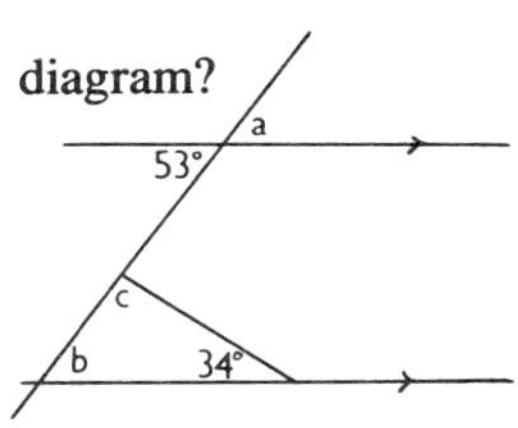

31. Circles

You will have to know what the following are:

(1) **Tangent.** This a line touching the circle on the outside. It is such that it cannot, if extended, cut the circle.

(2) **Chord.** This is a line inside the circle which cuts it into two parts called segments.

(3) **Segment.** This is one section of a circle which has been cut by a chord.

(4) **Arc.** This is a part of the circumference of a circle.

(5) **Sector.** This is a pie-like chunk of a circle.

(6) **Radius.** This is a line between the centre of the circle and the circumference.

(7) **Diameter.** This is the line that cuts the circle exactly in half. It passes through the centre of the circle and is therefore 2 radii in a straight line.

(8) **Circumference.** This is the perimeter of the circle.

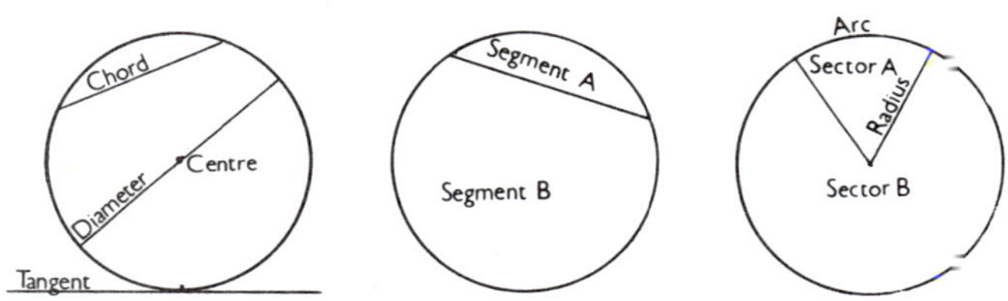

Circumference of a Circle.

RULE: $\mathbf{C = 2 \pi r}$

where C is the circumference or perimeter.
r is the radius.
π is 3.142, $\frac{22}{7}$ or as on your calculator.

Example. Find the circumference of a circle whose diameter is 14 cm. Take π as $\frac{22}{7}$

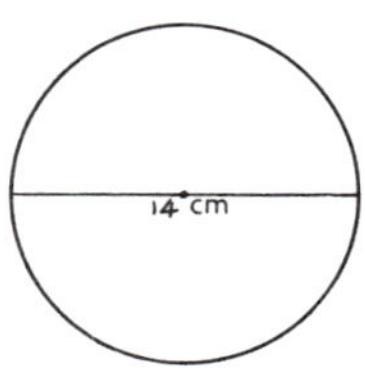

$$C = 2 \pi r$$

$$C = 2 \times \frac{22}{7} \times 7 = 44 \text{ cm}$$

Area of a Circle

***RULE:* Area = πr^2**

Example. Find the area of a circle whose perimeter is 132 feet.

Since $C = 2\pi r$, $\quad 132 = 2 \times \frac{22}{7} \times r$

change subject $\quad \frac{7 \times 132}{2 \times 22} = r$

$r = 21$ feet.

So if Area $= \pi r^2$ $\quad$ Area $= \frac{22}{7} \times 21 \times 21$

Area $= 1386$ square feet. (ft^2)

How do I remember which formula is for which?

***RULE:* Remember SQUARED IS FOR AREA.** Area is always given in say cm^2, m^2, etc so the formula with πr^2 is for area.

*REMEMBER that a semi circle is half of a circle.

Example. In the diagram of a paddling pool, the sides are straight and the end is in the form of a semi circle. What is the area of the pool and what is the perimeter of the pool?

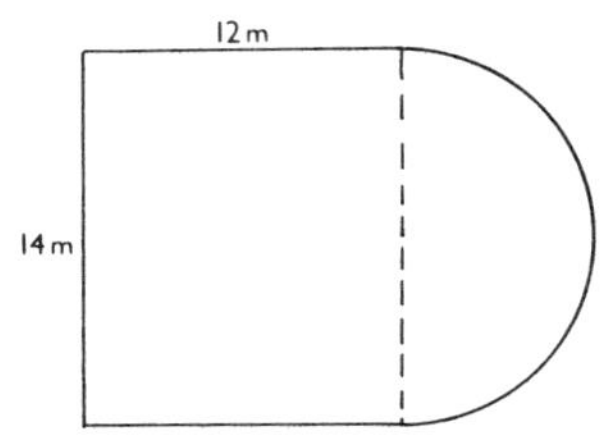

Area of rectangle is $14 \times 12 = 168\ m^2$

Area of semi circle is $\frac{\pi r^2}{2} = \frac{22}{7} \times \frac{7 \times 7}{2} = 77\ m^2$

Total area is $245\ m^2$

Circumference of semi circle is $\frac{2\pi r}{2} = \frac{2 \times 22 \times 7}{2 \times 7} = 22m$
(curved part only)

Total perimeter of pool is $12 + 14 + 12 + 22 = 60$ m

*REMEMBER for PERIMETERS imagine you are walking along each side of the figure. Put a mark where you begin and mark down the length of each side as you walk it. That way you will avoid adding in the diameter of the semi circle or omitting any bits.

SOME OTHER PROPERTIES OF CIRCLES

(1) **Angles in the same segment are equal.**
This is often said to be angles subtended by the same arc.

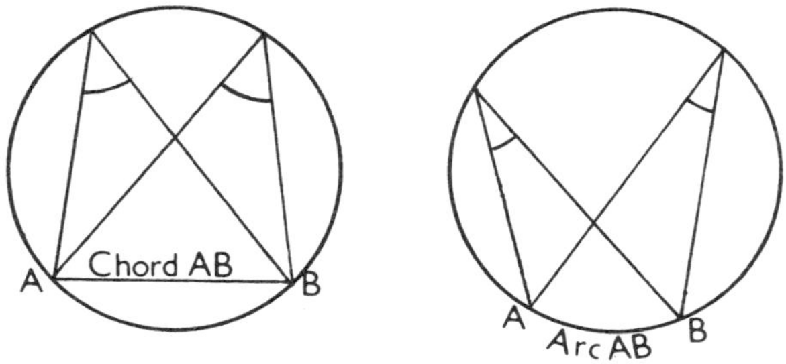

Beware of tricky questions where there are angles but no chords are actually drawn in. They are subtended by an arc. You can draw in your own chord connecting the ends of the angles.
LEARN to recognize the two triangles apex to apex (top to top).

(2) **Angle in a semi circle is 90°.**

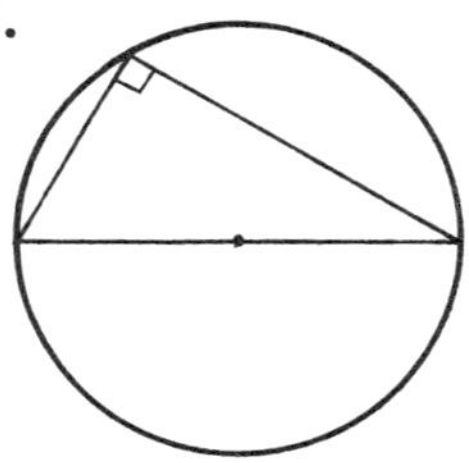

(3) **Angle between tangent and a radius is 90°.**
Radius always meets tangent at a right angle.

Remember that some questions on right angled triangles may involve PYTHAGORAS.

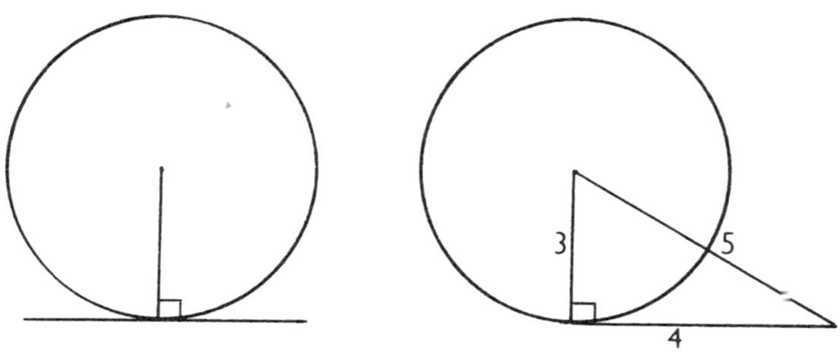

(4) **Radius at right angles to a chord bisects that chord.**

AX = AB

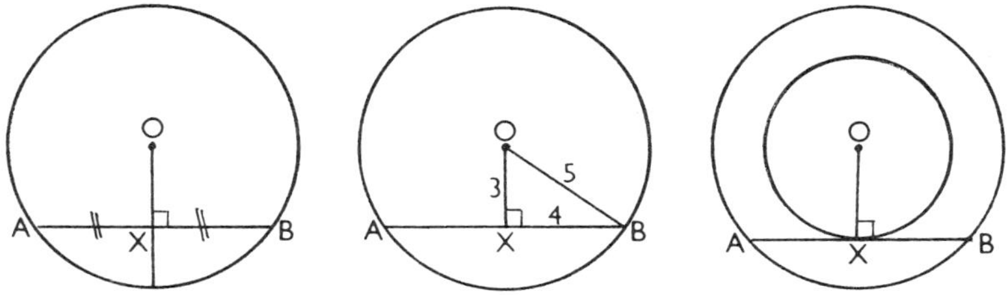

(5) **Radii of a circle are equal in length so they will make isosceles triangles.**

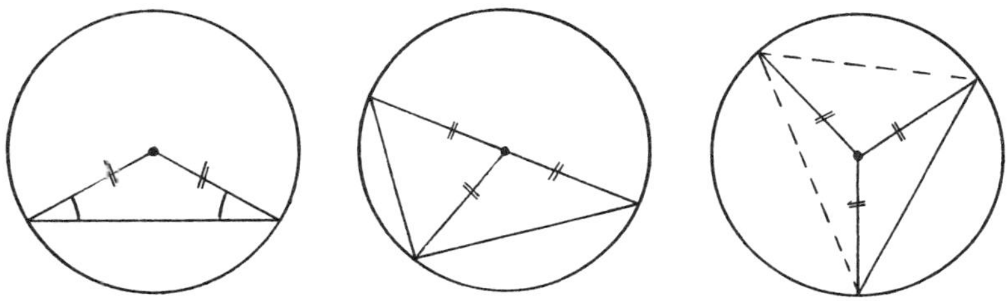

Some Examples of the use of these Rules.

Example. In the figure (a) below, a semi circular bowl of diameter 10 cm is filled with water to a depth of 2 cm. What is the diameter of the surface of the water?

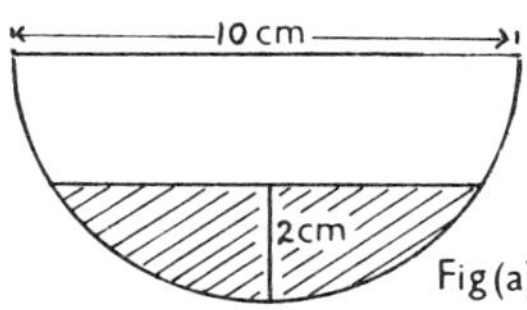

Fig(a)

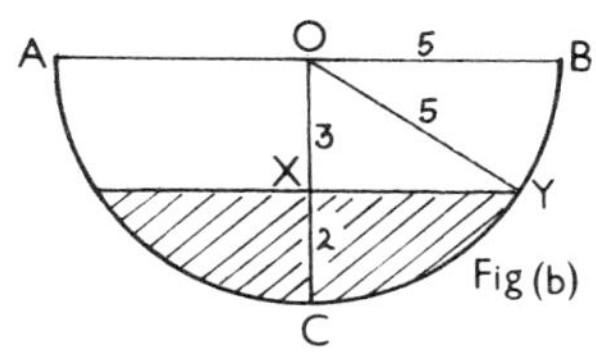

Fig (b)

In fig (b) letter the diagram and XY is the radius of surface.

Since the radii of a circle are all even in length, we can draw in radius OY.

and since OB is half the diameter, it is 5 cm.

so OY = 5 cm

Since OC is 5 cm, then OX = 3 cm

XY is 4cm. OXY is a 3, 4, 5 triangle using PYTHAGORAS.

Therefore diameter of surface of water is 8 cm.

Example. In the figure below find the angles of the triangle ABC.

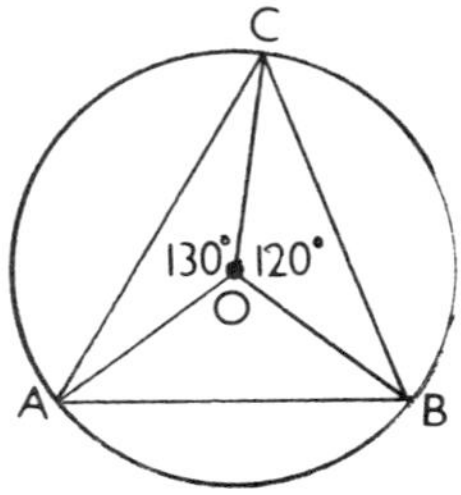

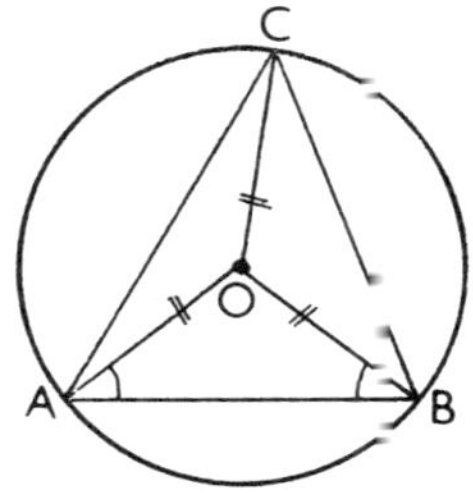

Angle AOB = 110° Angles at a point

Radii are equal in length so we know the base angles of triangles formed are equal.

so Angle OBC = OCB = 30°
and Angle OCA = OAC = 25°
and Angle OAB = OBA = 35°

So the angles of the triangle ABC are 30° + 25° = 55°
25° + 35° = 60°
and 35° + 30° = 65°

EXERCISE N.

(1) In the diagram below find angles marked a, b, c and d.

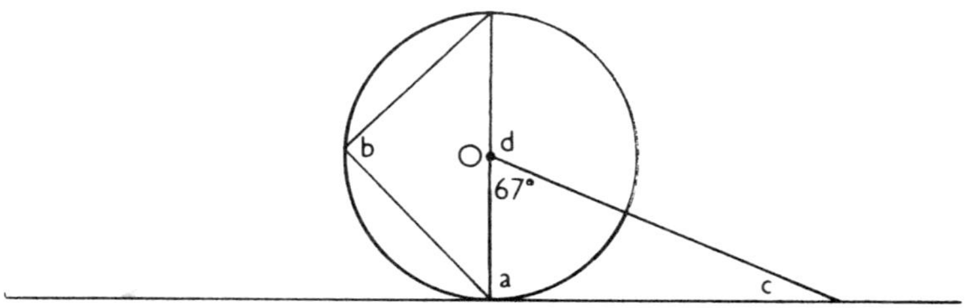

(2) In the diagram below, calculate angles a, b, and c and the length of CX.

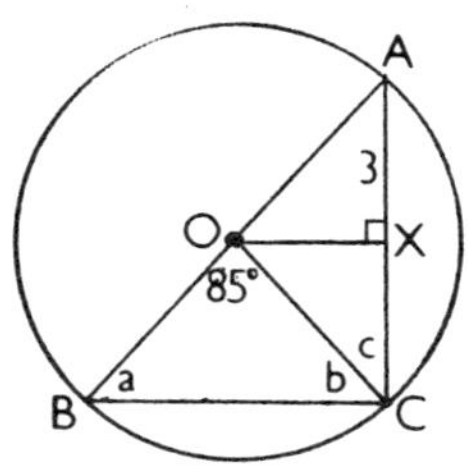

32. Properties of Quadrilaterals

A quadrilateral is a four sided figure, just as a triangle is a three sided figure. And just as we can have varying forms of triangle, so we can have different types of quadrilateral. We'll start with the best known first. With each type is a list of the things we know WITHOUT ANY DOUBT about the figure.

(1) **The Square.**

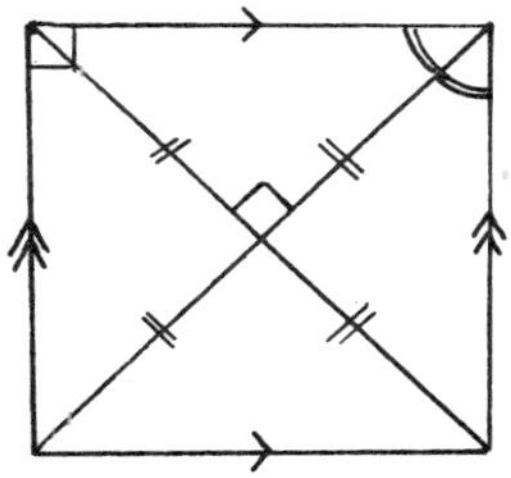

All sides are equal length
Opposite sides are parallel
All angles are equal and are 90°
Diagonals are equal in length
Diagonals bisect the corner angle
Diagonals bisect each other
Diagonals cross at right angles.

Area = breadth × height

(2) **Rectangle.**

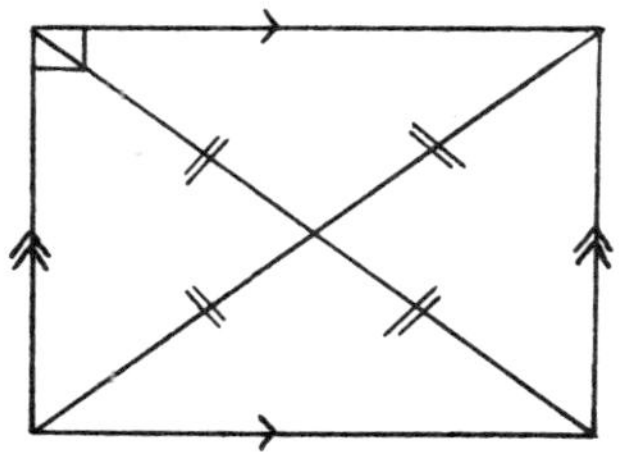

Opposite sides are equal length
Opposite sides are parallel
All angles are right angles (90°)
Diagonals are the same length
Diagonals bisect each other.

Area = breadth × height

(3) **Rhombus.**

A SQUASHED BOX or squashed square making a DIAMOND shape. Think of a square that has been tilted. A rhombus has almost all the properties of a square.

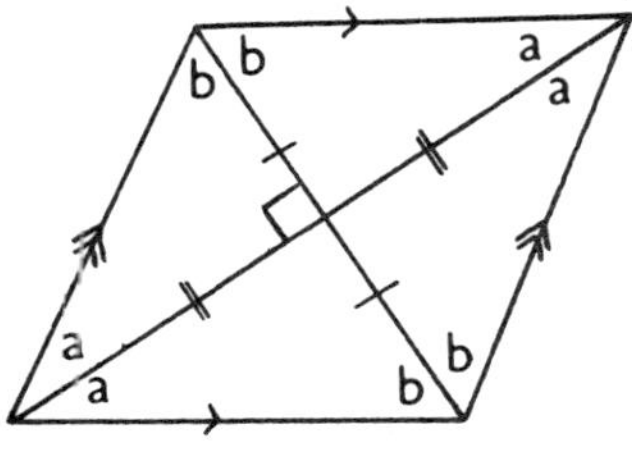

All sides are equal length
Opposite sides are parallel
Opposite angles are equal
Diagonals bisect the corner angle
Diagonals bisect each other
Diagonals cross AT RIGHT ANGLES

Area = breadth × vertical height

(4) **Trapezium.** AN ELEPHANT BOX.

Many people get mixed up between a trapezium and a rhombus. The way I always remember what a trapezium looks like is as follows,

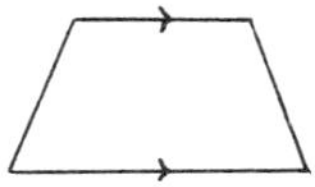

trapezes are used in a circus
circuses have elephants
elephants stand on elephant boxes
And we know what they look like.

Opposite sides are parallel

Area = average breadth × vertical height

The two non-parallel sides can be different lengths and may stick in or out.

(5) **Parallelogram.** A SQUASHED RECTANGLE.

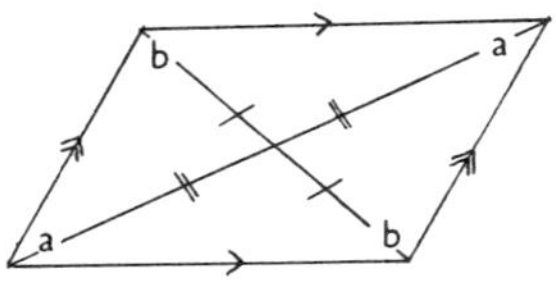

Opposite sides are parallel
Opposite sides are equal length
Opposite angles are equal
Diagonals bisect each other.

Area = breadth × vertical height

***RULE:* INTERNAL ANGLES OF *ANY* QUADRILATERAL ADD UP TO 360°**

These rules are important in that they help us in our calculations of angles and areas.

Example. In the rhombus below, calculate the angles a, b, and c.

Since we know it's a rhombus we can apply its properties. Add in angle f and g

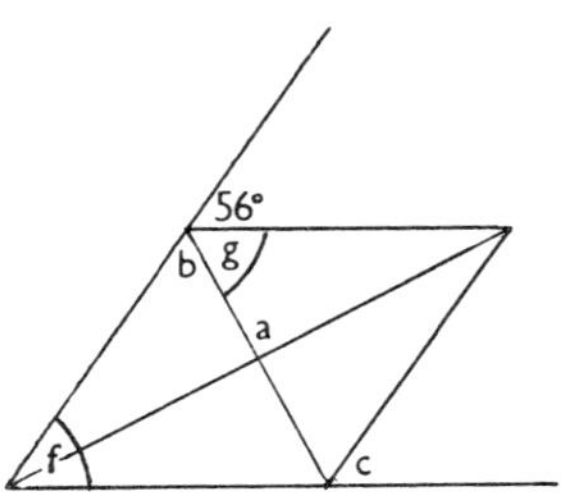

a = 90°. Diagonals cross at right angles.

c = f = 56° Corresponding angles.

b = g so 2b + 56° = 180°
2b = 124°
b = 62°

33. Polygons

POLY means MANY. A polygon has many sides. Anything that is not a triangle or a quadrilateral is a polygon.
REGULAR polygons have sides of equal length and therefore the interior angles are equal.

Five Sides | Six Sides | Seven Sides | Eight Sides

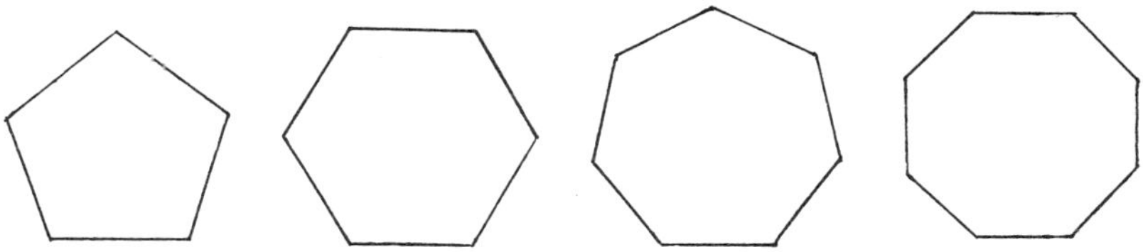

Pentagon | Hexagon | Heptagon | Octagon

How do we calculate the INTERIOR angle of ANY regular polygon?

REMEMBER that an EXTERIOR angle is an angle made when we extend one of the sides. It is NOT the reflex angle.

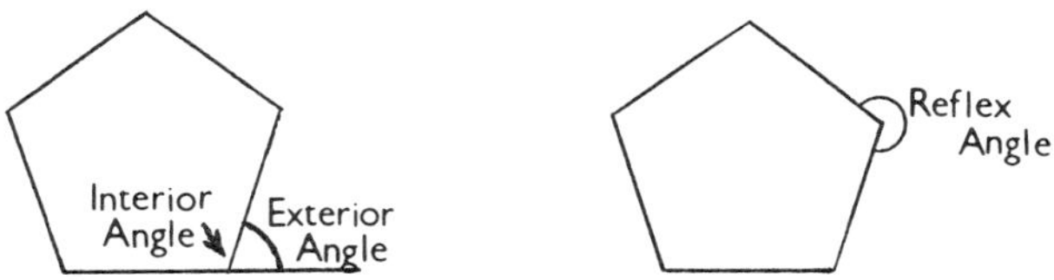

So if we know the size of the EXTERIOR ANGLE, then we can work out the size of the INTERIOR ANGLE. (Angles on a straight line).

***RULE:* ALL THE *EXTERIOR* ANGLES OF A POLYGON ADD UP TO 360°.**

Example. Find the size of the interior angle of a regular eight sided figure.

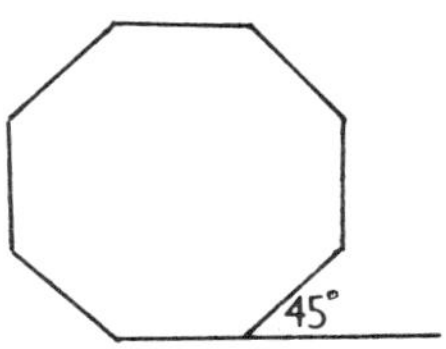

Exterior angles add up to 360°
There are eight exterior angles
So each EXTERIOR angle is

$$\frac{360}{8} = 45°$$

INTERIOR angle is 180° − 45 ° = 135°

Example. How many sides has a polygon whose internal angles are 150°

Interior angle is 150°
Exterior angle is 30°

$$\frac{360}{30} = 12$$

The Polygon has 12 sides.

34. Areas and Volumes

(1) Square and Rectangle.

A square measuring 1 cm × 1 cm is 1 SQUARE CM. (1 cm^2)

A square measuring 2 cm × 2 cm is 4 SQUARE CM. (4 cm^2)
It has 2 rows of 2 cm.

2 cm × 2 cm

A rectangle measuring 2 cm by 3 cm is 6 SQUARE CM. (6 cm^2)
It has 2 rows of 3 cm.

2 cm × 3 cm

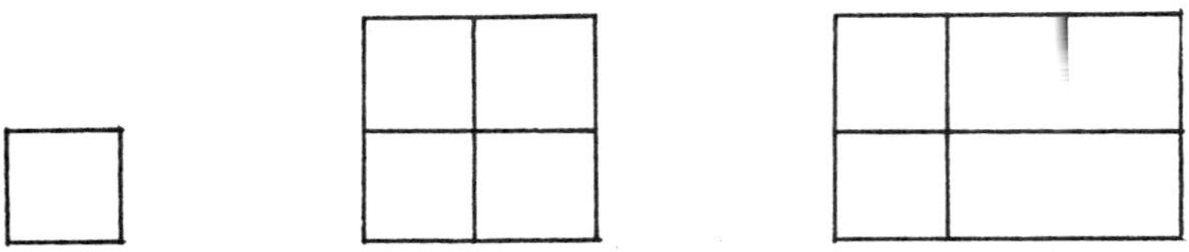

1 cm^2 4 cm^2 6 cm^2

***RULE:* Area = breadth times height. A = b × h**

(2) Cube and Cuboid.

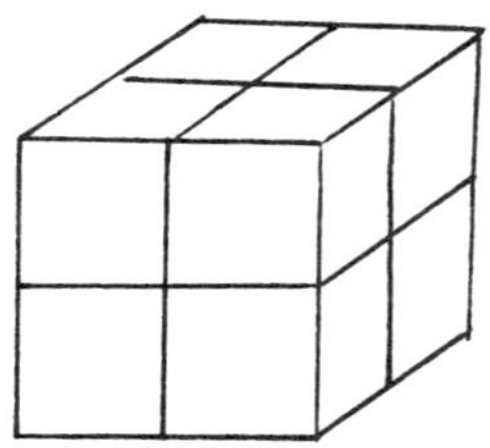

A cube measuring 1 cm × 1 cm × 1 cm is 1 CUBIC CM (1 cm^3)

A cube measuring 2 cm × 2 cm × 2 cm is 8 CUBIC CM (8 cm^3)

It is a block of 8 little blocks measuring 1 cm × 1 cm × 1 cm

***RULE:* Volume = breadth × height × Length. V = b × h × L**

BUT b × h is the AREA

***RULE:* Volume = Area × Length. V = Area × L**

THIS RULE APPLIES TO ANY SHAPE SO LONG AS IT IS THE SAME SHAPE THROUGHOUT ITS ENTIRE LENGTH.

Something that is the same shape throughout its entire length is said to be *OF UNIFORM CROSS SECTION*

(3) **Triangle**

***RULE:* Area of a triangle is $\frac{1}{2}$ × breadth × height.**

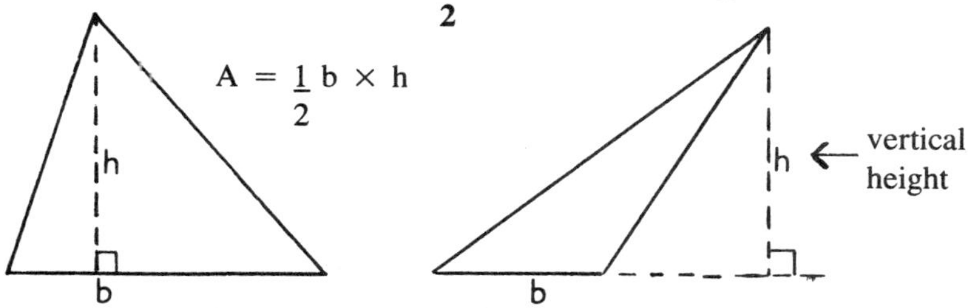

Acute angled Obtuse angled

$$A = \frac{1}{2} b \times h$$

(4) Triangular PRISM

A triangular prism is formed when a solid has a triangular *UNIFORM CROSS SECTION*

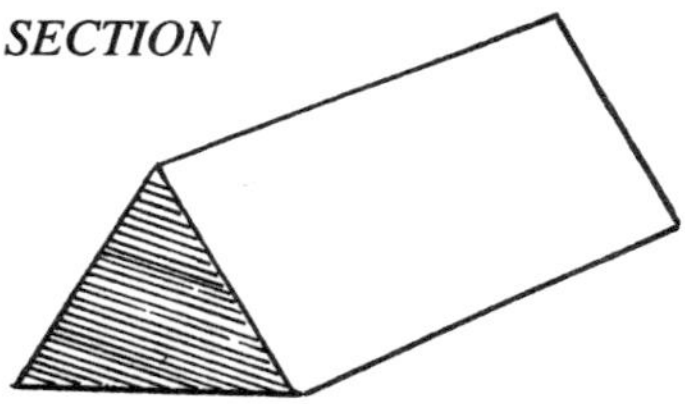

$$V = \text{Area} \times L$$

(5) Parallelogram.

***RULE:* Area of a parallelogram is breadth × *vertical* height.**

$$A = b \times h$$

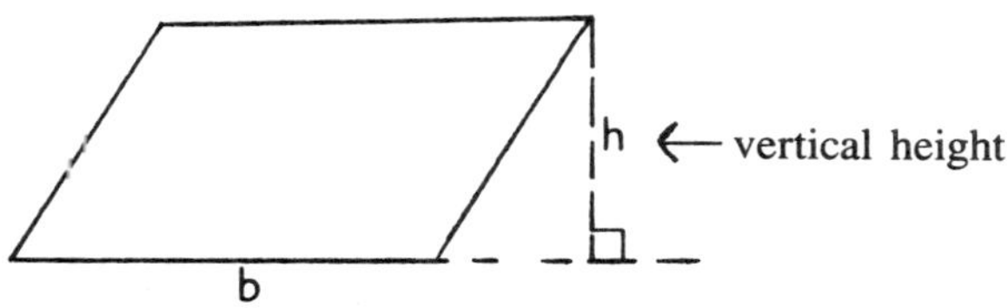

(6) Trapezium.

Since the breadth is neither that of the top or the bottom, we take an AVERAGE.

***RULE:* Area of a trapezium is AVERAGE breadth × vertical height**

$$A = \frac{a + b}{2} \times h$$ where a and b are the top and the bottom

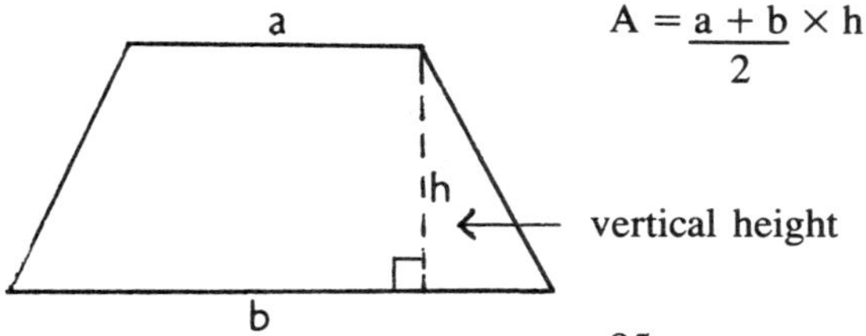

(7) Rhombus.

Area of a rhombus is breadth × vertical height.

$$A = b \times h$$

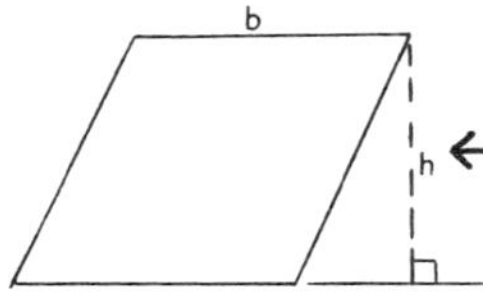

← vertical height.

BUT, given the lengths of the diagonals we can still work out the area of a rhombus.

Example. Given the lengths of the diagonal of a rhombus are 12 cm and 8 cm, what is its area?

Since we know that the diagonals cross at right angles, we have 2 triangles of which we know the breadth and the vertical height.

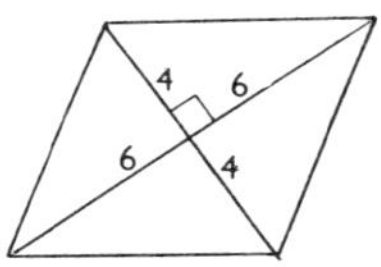

Area of triangle A $= \frac{1}{2} \times 12 \times 4$

$= 24\ cm^2$

Area of triangle B is also $24\ cm^2$

so Area of rhombus is $48\ cm^2$

(8) Cylinder.

A CYLINDER is of circular *UNIFORM CROSS SECTION.*

***RULE:* VOLUME OF A CYLINDER is area of the circular base × length**

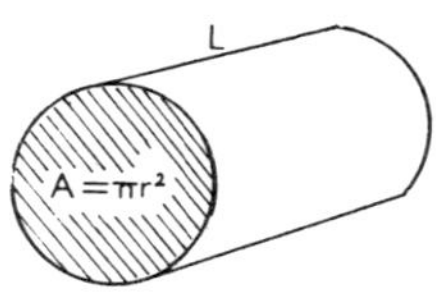

$$V = \pi r^2 \times \text{Length}$$

Area of the curved surface of a cylinder is the same as opening it out into a rectangle. Think of rolling up a sheet of paper into a tube, and the breadth of the paper becomes the outer edge of the cylinder.

***RULE:* AREA OF A CYLINDER is the circumference × height.**

$$A = 2\pi r \times \text{height.}$$

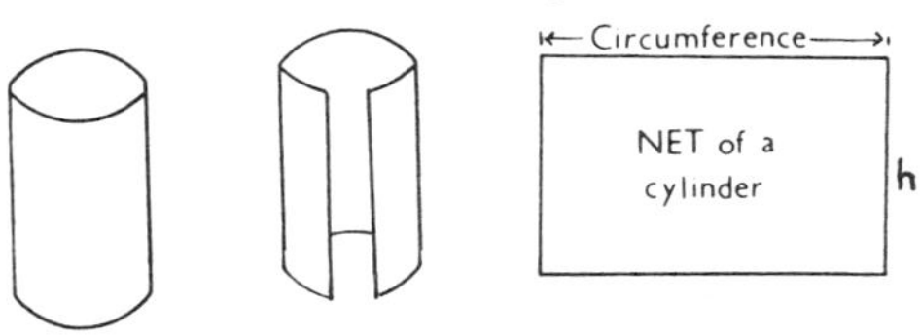

(9) Other Figures.

Areas of other shapes are worked out using one or more of the above rules.

Example. The diagram shows the uniform cross section of a dam. If the dam is 400 m long what volume of concrete went into making it?

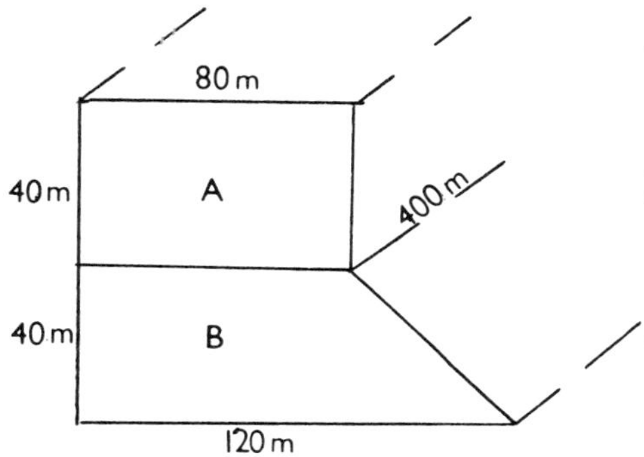

$$\text{Area of A} = b \times h = 80 \times 40 = 3200 \text{ m}^2$$

Area of B

$$\text{Trapezium} = \frac{80 + 120}{2} \times 40 = 4000 \text{ m}^2$$

$$\text{Total area} = 7200 \text{ m}^2$$

$$\text{Volume} = \text{Area} \times \text{Length} = 7200 \times 400 = 2{,}880{,}000 \text{ m}^3$$

Example. The diagram shows in centimetres the cross section of a girder whose total volume is 24,000 cm^3. The weight of 1 cubic centimetre of metal is 7.6 grams. What is

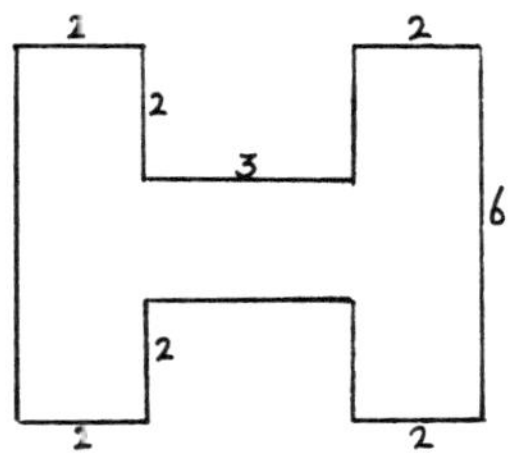

(a) the area of the cross section?
(b) the weight of the girder in kg?
(c) the length of the girder in metres?

(a) Area of cross section is TOTAL area LESS the missing bits.

Total area = 42 cm^2
Area of one bit is $2 \times 3 = 6$ cm^2
Area of other bit is $2 \times 3 = 6$ cm^2

Area of uniform cross section is $42 - 12 = 30$ cm^2

*NOTICE it is easier to work out the whole area of a rectangle and take the bits away than splitting the cross section up into smaller rectangles.

(b) If the volume of the girder is 24,000 cm^3
Then the weight is 24,000 × the weight of 1 cm^3

REMEMBER, if you have difficulty use THE RULE OF X

$1\ cm^3 = 7.6$ grams
$24{,}000\ cm^3 = X$ grams

$$24{,}000 \times 7.6 = 182{,}400 \text{ grams}$$

AND since

1000 grams = 1 kg
182,400 grams = X kg

$$\frac{182{,}400}{1000} = 182.4 \text{ kg weight of girder}$$

(c) Since $V = \text{Area} \times \text{Length}$

we can change the subject

$$L = \frac{V}{\text{Area}} = \frac{24{,}000}{30} = 800 \text{ cm}$$

Length of girder is 8 metres.

Example. The diagram shows the uniform cross section of a gun barrel whose length is 6 metres. Calculate the volume of metal.

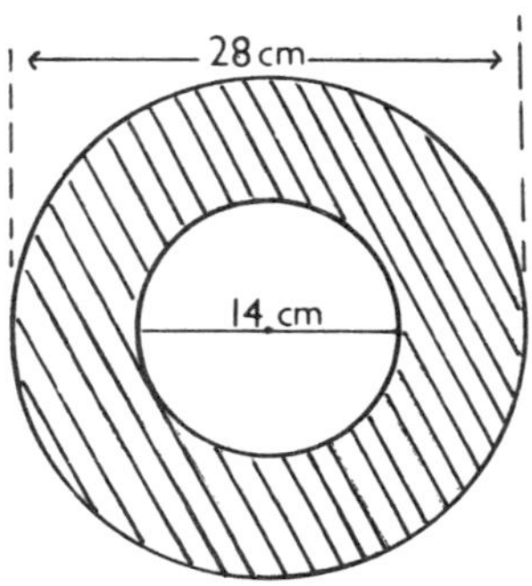

Total Area $= \pi r^2$
$= \frac{22}{7} \times 14 \times 14$
$= 616\ cm^2$

Area of hole $= \pi r^2$
$= \frac{22}{7} \times 7 \times 7$
$= 154\ cm^2$

Area of cross section of metal $= 616 - 154$
$= 462\ cm^2$

Volume = Area × Length

so Volume $= 462 \times 600 = 277{,}200\ cm^3$ of metal

NOTICE we converted the length to cm from metres.
REMEMBER the rule **DON'T MIX CM AND METRES IN AREAS AND VOLUMES**

(10) Other Solids

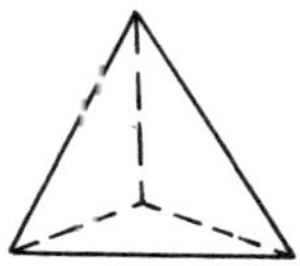

Tetrahedron.
A triangular based pyramid.

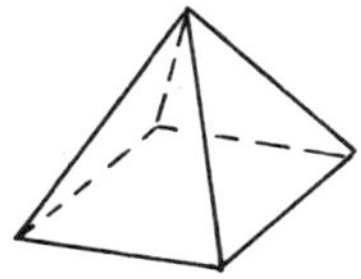

Square based pyramid.

A word about PRISMS

Prisms can have a cross section with 3 or more sides. A hexagonial prism has a cross section of 6 sides, a trapezoid prism has a cross section of a trapezium. And so on.

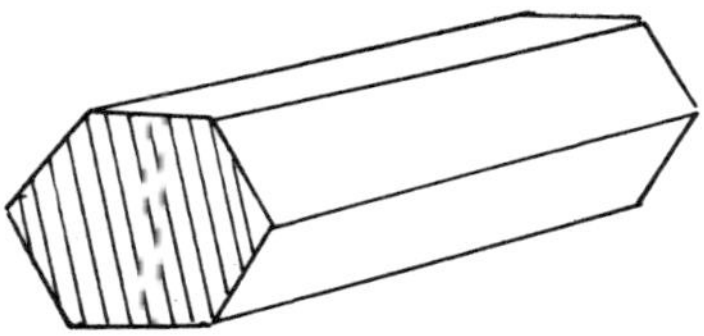

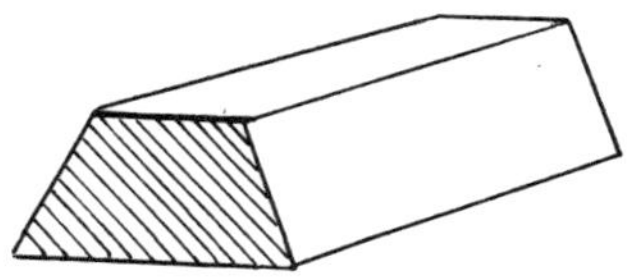

35. Nets

A NET is a paper pattern. When it is folded it makes a solid figure.

A net of a rectangular box looks like this

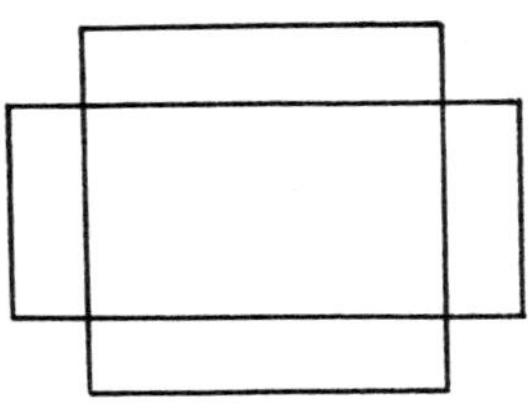

A net of a square based pyramid looks like this

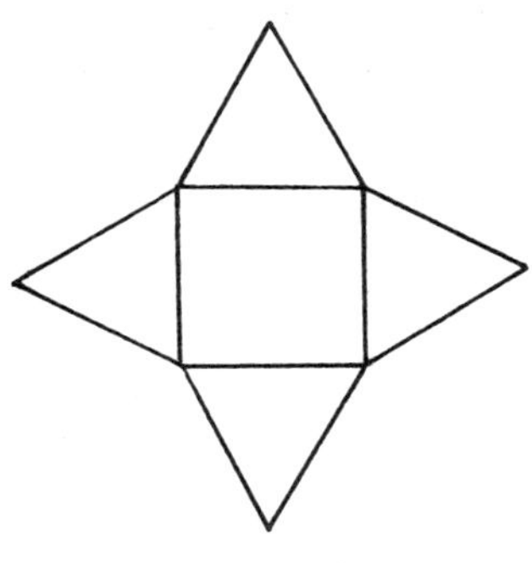

36. Tessellation

***RULE:* TESSELLATION MEANS FITTING TOGETHER WITHOUT ANY GAPS.**

At every point of contact the angles add up to 360°

Common examples of figures that fit together are:

Triangles

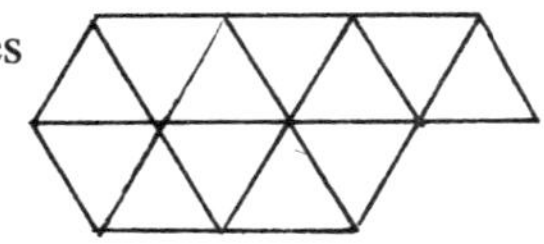

Squares

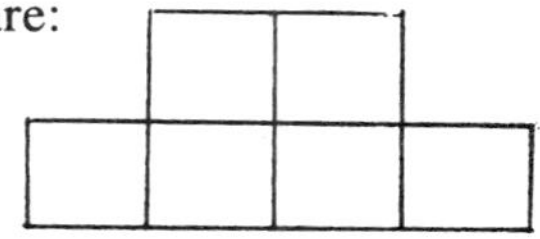

Hexagons

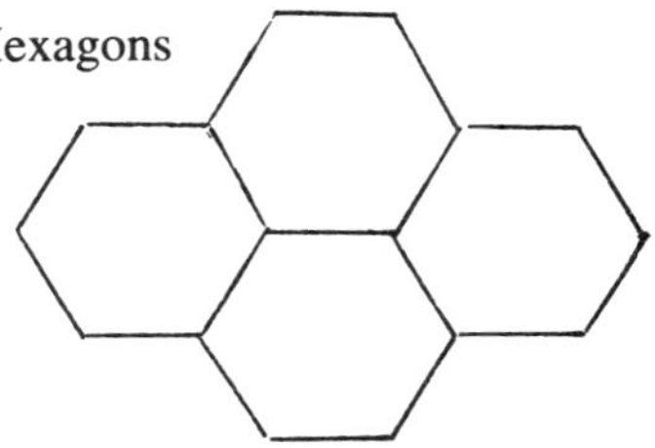

Rhombs

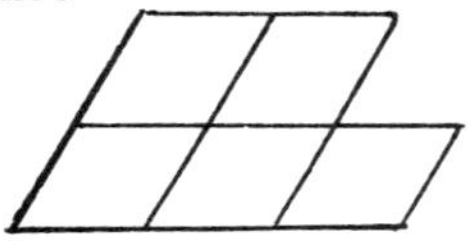

EXERCISE O.

(1) Draw the net of a tetrahedron.

(2) What is the volume of a cuboid measuring 6.5 cm × 12 cm × 5 cm

(3) What is the size of the interior angle of a regular pentagon?

(4) What is the difference between a trapezium and a rhombus?

(5) A trapezium and a rhombus are both 5 cm high. If the length of the rhombus is 6 cm, and the top of the trapezium is 4 cm, what is the length of the base of the trapezium if the two figures have the same area?

(6) How do we know that figures will tessellate?

(7) The diagram shows the uniform cross section of a metal bar (in cm).

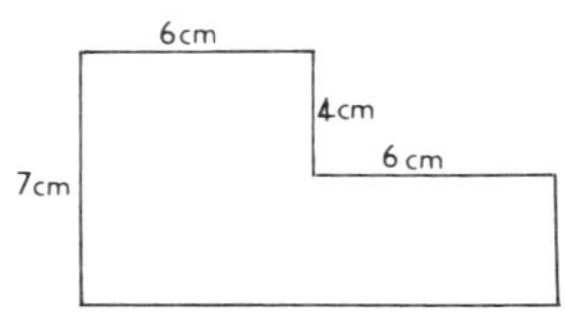

(a) What is the area of the cross section?

(b) What is the length of the bar in metres if the volume of the bar is 25,500 cm^3?

(c) What is the weight of the bar if a 5 cm cube of the metal weighs 0.925 kg?

(d) What is the weight of 1 cm^3 of the metal?

(8) Find the internal angle of a regular ten sided figure.

37. Congruent Triangles

CONGRUENT figures are the same shape and the same size. They are the same in all respects.

SIMILAR figures are the same shape but a DIFFERENT size.

A triangle is congruent to (same as) another triangle when

(1) All its sides are equal in length to the sides of the other triangle. (Side Side Side) SSS.

(2) Two sides and the angle between those two sides are the same as the other triangle. (Side Angle Side) SAS.

(3) Two angles and then a side are equal to those of the other triangle. (Angle Angle Side) AAS.

(4) The triangles are both right angled and the hypotenuse and a side are equal to those of the other triangle. (Right angle Hypotenuse Side) RHS.

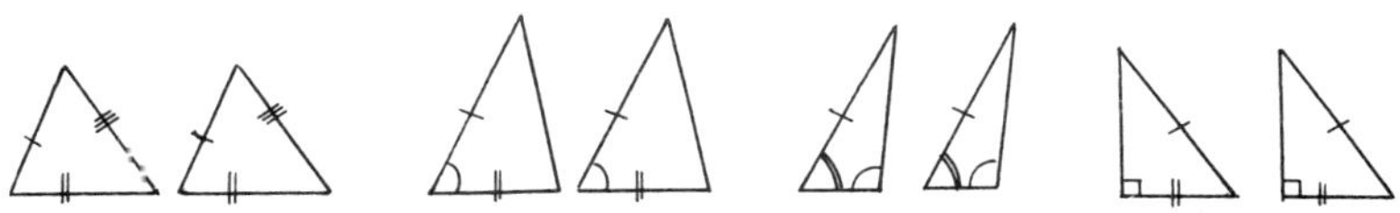

SSS SAS AAS RHS

LEARN these because other combinations may not be congruent. For instance, if the triangles are ASS (Angle Side Side, then one could be acute angled and the other obtuse angled.

Try constructing a triangle with Angle 45°, Side 5 cm, Side 4 cm.

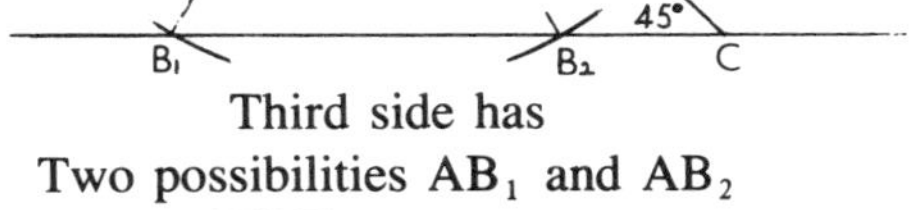

Third side has
Two possibilities AB_1 and AB_2
NOT congruent

REFLECTED triangles are congruent.
ROTATED triangles are congruent.
TRANSLATED triangles are congruent.

38. Transformations

(1) Reflections.

A reflection is a mirror image. Triangle Y is a reflection of triangle X.

The reflection takes place ACROSS a line.

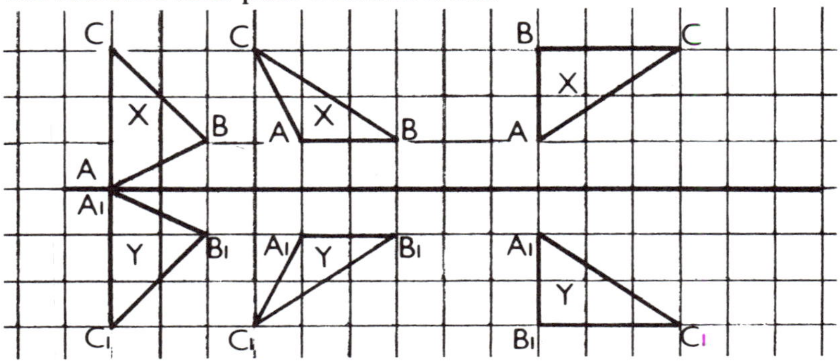

If A touches the line then A_1 touches the line.
If B is 1 cm from the line then B_1 is 1 cm from the line.
If C is 3 cm from the line then C_1 is 3 cm from the line.
and so on.

***RULE:* EACH POINT IS REFLECTED. TRANSFER EACH POINT SEPARATELY.**

A good test of reflection is if you can fold along the line and the images fall exactly together then it is a true reflection.

The line about which the reflection takes place is

A LINE OF SYMMETRY

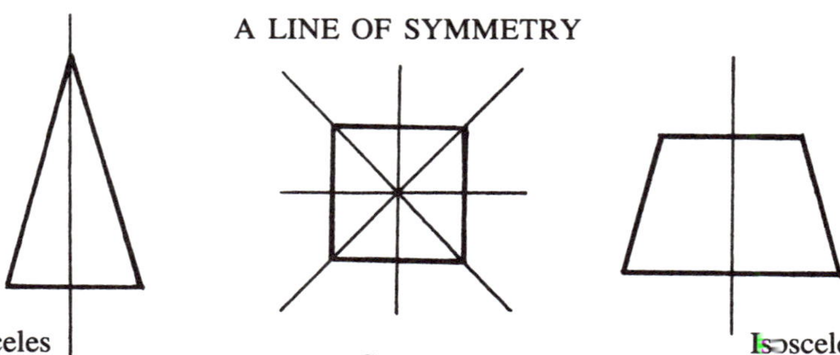

Isosceles Triangle

Square

Isosceles Trapezium

Lines of Symmetry can be drawn through any figure that has identical halves. As you can see, some figures have more than one line of symmetry.

(2) Rotations.

Rotations take place about a POINT rather than a line.
Usually we are asked to rotate a figure 180° or 90° clockwise, or 90° anti-clockwise (90° anti-clockwise is 270° clockwise)

***RULE:* EACH POINT IS ROTATED. ROTATE EACH POINT SEPARATELY.**

***RULE:* ROTATE THOSE LINES THAT ARE AT RIGHT ANGLES TO THE AXES FIRST.**

***RULE:* IF IN DOUBT USE COMPASSES ON EACH POINT.**

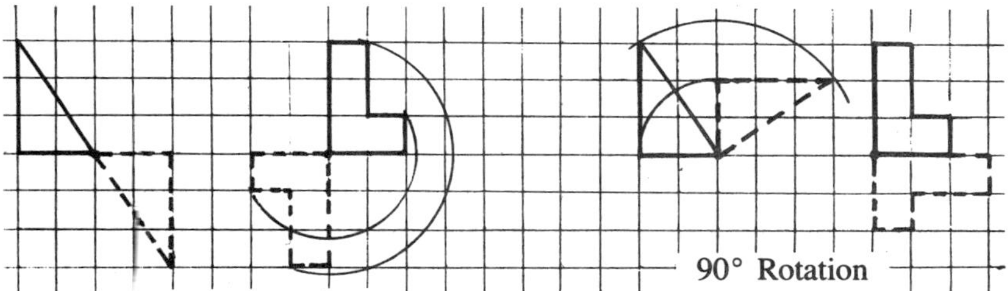

(3) Translations.

A figure is TRANSLATED if it is moved from one place to another. Graphically this is done with VECTORS.

***RULE:* EACH POINT IS TRANSLATED. MOVE EACH POINT SEPARATELY.**

If I move a triangle 4 squares on and 1 square up I have translated each point 4 on and 1 up.
This is a vector $\begin{pmatrix} 4 \\ 1 \end{pmatrix}$ Point A moves to A_1 point B to B_1 point C to C_1

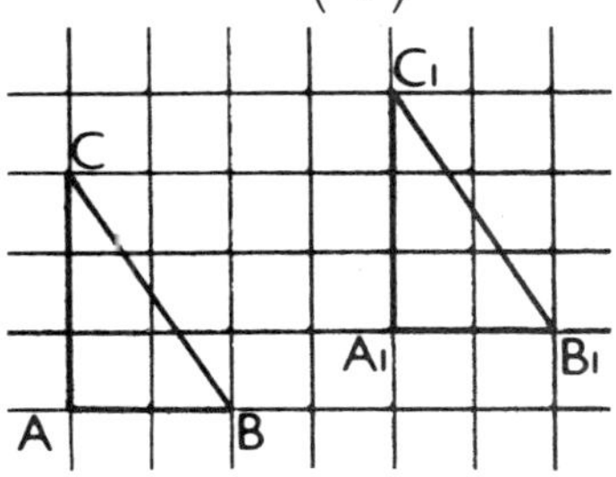

Vector $\begin{pmatrix} 4 \\ 1 \end{pmatrix}$

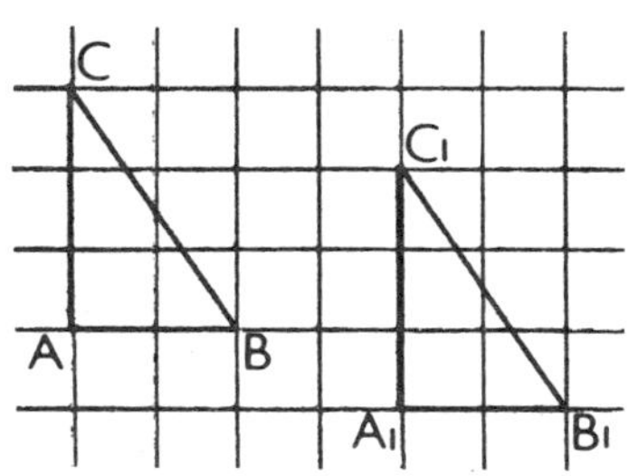

Vector $\begin{pmatrix} 4 \\ -1 \end{pmatrix}$

Enlargement.

A fourth type of Transformation is an ENLARGEMENT. This is a scaled up version of the original figure. These figures are not congruent, but they are SIMILAR. They are scaled up by SCALE FACTORS.
For a scale factor of 2 we double the length of each side of the figure.

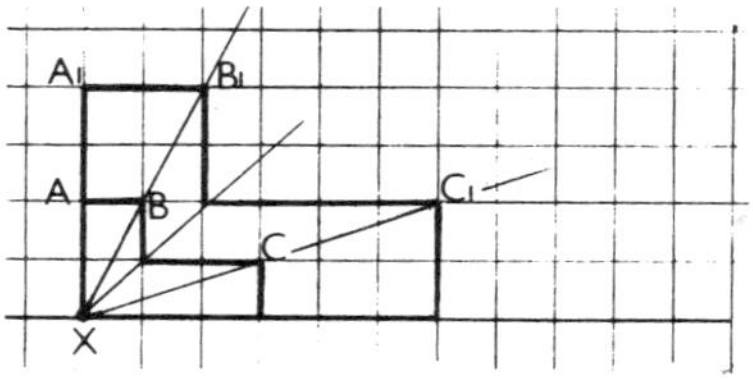

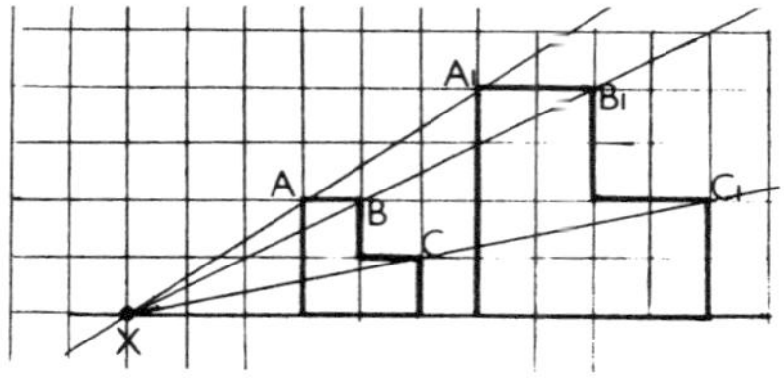

Centre of Enlargement or Reduction.

This is the point through which similar points on the two figures can be connected.
Lines through A_1 and A meet lines through B_1 and B and through C_1 and C etc. at X.

Reductions.

A scale factor of $\frac{1}{2}$ would reduce a figure to half its size.

EXERCISE P.

(1) Are the following triangles below congruent?

a.

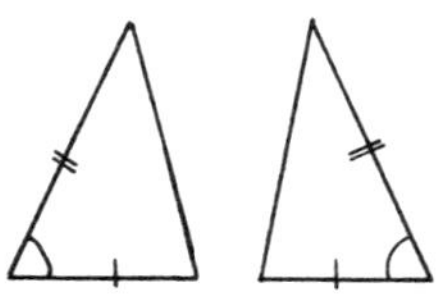

b.

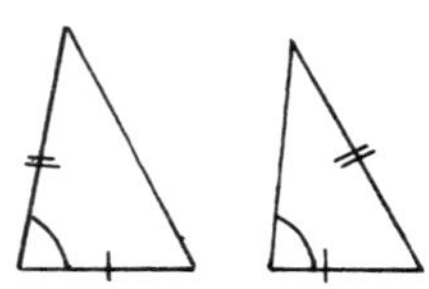

c.

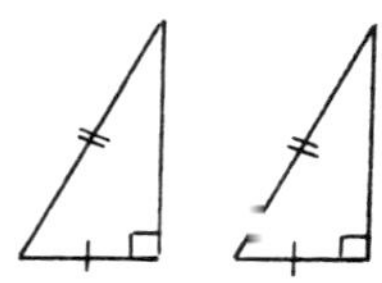

(2) In the diagram, what is the scale factor of the enlargement?

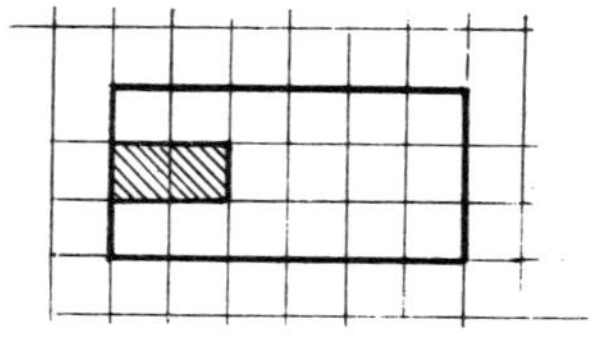

(3) In the diagram, what transformation maps triangle A:
(a) onto triangle B, (b) onto triangle C, (c) onto triangle D?

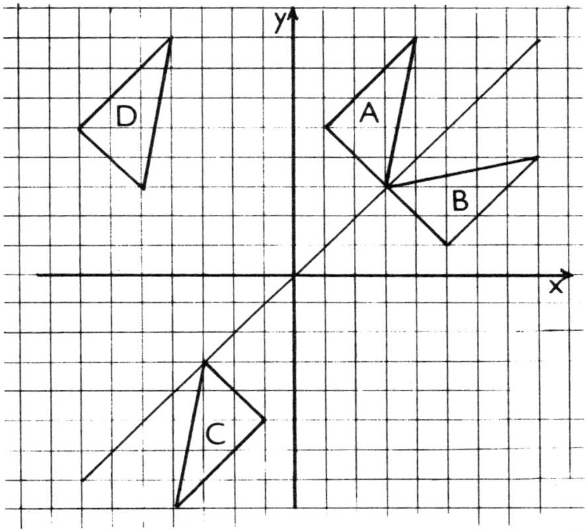

39. Trigonometry

CALCULATIONS IN RIGHT ANGLED TRIANGLES

In the triangle, one angle is marked *a*

Notice the hypotenuse is the side opposite the right angle.

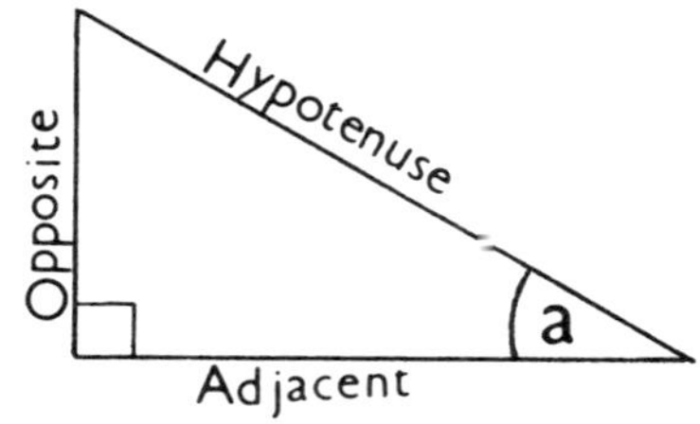

Of the other two sides, one is opposite *a* and the other is adjacent to it.

So we have three sides of a right angled triangle

Hypotenuse	H
Opposite	O
Adjacent	A

For calculations in trigonometry learn the following rule:

***RULE:* SOH CAH TOA.**

$$\text{Sin } a = \frac{\text{Opposite}}{\text{Hypotenuse}} \qquad S = \frac{O}{H}$$

$$\text{Cos } a = \frac{\text{Adjacent}}{\text{Hypotenuse}} \qquad C = \frac{A}{H}$$

$$\text{Tan } a = \frac{\text{Opposite}}{\text{Adjacent}} \qquad T = \frac{O}{A}$$

REMEMBER: SOH CAH TOA. It sounds like an Hawiian volcano
so-ka-to-a. Say it to yourself until you know it

That's the hardest bit over. The rest is just substitution of values in the three formulas above.

Example. Find the length of side BC

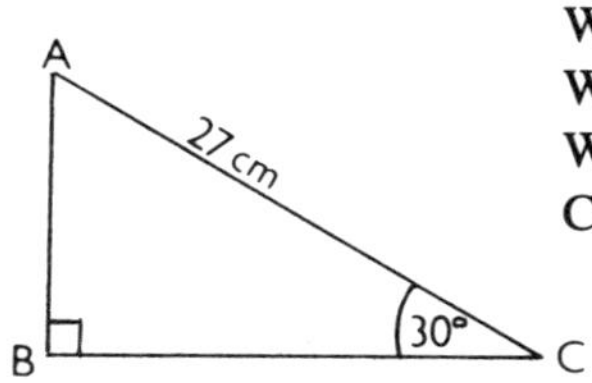

We have the angle 30°
We have the hypotenuse = 27 cm
We want the Adjacent side.

CAH $\qquad \text{Cos} = \frac{\text{Adj}}{\text{Hyp}}$

$$\text{Cos } 30° = \frac{\text{Adj}}{27}$$

Change subject
Bottom goes to Top $\quad 27 \times \cos 30° = \text{Adj}$

Use the calculator. Enter 27 × 30 cos =

23.382686 will appear on the display.
BC = 23.38 cm (to 2 decimal places).

***RULE:* ENTER THE ANGLE FIRST AND THEN THE TRIG FUNCTION.**

So $\quad$ tan 24° $\quad$ enter as 24 tan

and $\quad$ sin 23° $\quad$ enter as 23 sin and so on.

Example. In the triangle find AB.

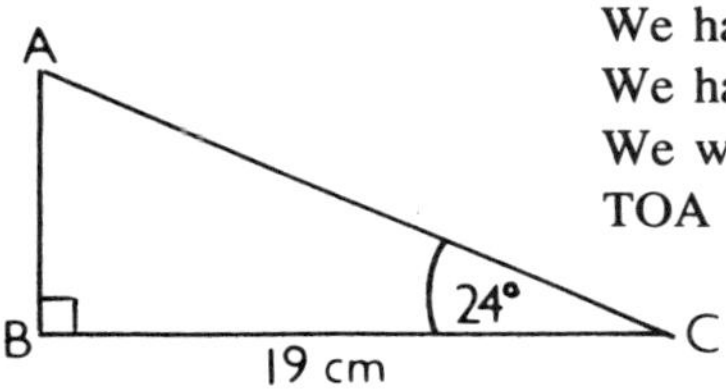

We have the angle 24°
We have the Adjacent side = 19 cm
We want the Opposite side.
TOA $\quad \text{Tan} = \dfrac{\text{Opp}}{\text{Adj}}$

$$\text{Tan } 24° = \frac{\text{Opp}}{19}$$

Change subject
Bottom to Top $\quad 19 \times \tan 24° = \text{Opp.} \quad = 8.459$

AB = 8.50 cm (to two decimal places).

Example. In the triangle, find AC

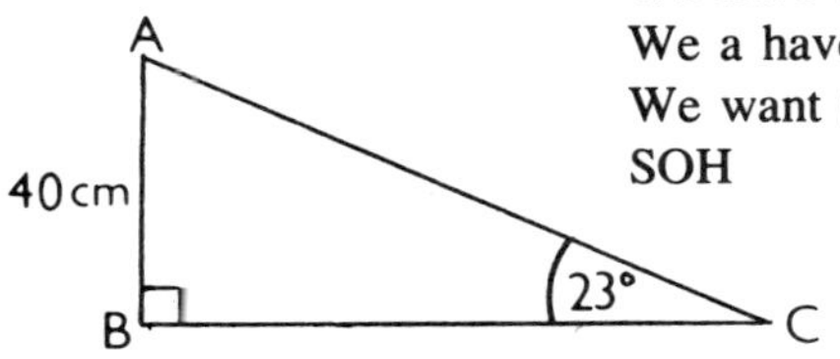

We have the angle 23°
We a have the Opposite side = 40 cm
We want the Hypotenuse
SOH $\quad \text{Sin} = \dfrac{\text{Opp}}{\text{Hyp}}$

$$\text{Sin } 23° = \frac{40}{\text{Hyp}}$$

Change subject
Top to Bottom
Bottom to Top $\quad \text{Hyp} = \dfrac{40}{\text{Sin } 23°}$

40 ÷ 23 sin = 102.372

AC = 102.37 cm (to 2 decimal places).

FINDING THE ANGLE.

Given two of the sides we can work back to find out the angle.

Example. Find the angle marked a in the triangle below.

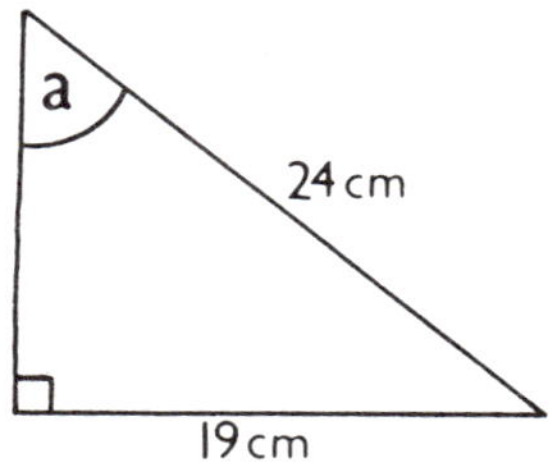

We have the Opposite side = 19 cm
We have the Hypotenuse = 24 cm

SOH $\text{Sin} = \frac{\text{Opp}}{\text{Hyp}}$

$\text{Sin } a = \frac{19}{24}$ Enter **19** **÷** **24**

Display shows Sin a = 0.7916667.

That is *SIN a.* It is what we get AFTER we have used Sin on an angle. So we have to invert the number to get back to *a.*

Press **INV** **sin** and we get 52.341541°

Press **DMS** or **°′″** and we get 52° 20′ 30″. Size of angle *a*
(which is in Degrees Minutes and Seconds).

(If we Sin this angle we end up with 0.7916667 so you can see why we have to reverse the sequence to get the angle).
Try Sin 50° and then get it back to 50°

***RULE:* USE INV TO CALCULATE THE ANGLE.**

***RULE:* USE A CALCULATOR WITH DMS OR °′″ KEY.**

Should your calculator not have the **DMS** or **°′″** button, then we know we have 52° but want to find what 0.341541 of a degree is.

Subtract 52° to leave the decimal fraction on display

THE RULE OF X.
There are 60 min in 1° gives 0.341541 × 60 Min.
X min in 0.341541
= 20.492463 min

Subtract 20 leaving the fraction of minutes on display

THE RULE OF X.
There are 60 seconds = 1 Min gives 0.492463 × 60
X seconds = 0.492463 Min
= 30 Sec.

In other words multiply the decimal fractions by 60.

The Isosceles Triangle.

REMEMBER that isosceles and equilateral triangles can be split into right angled triangles. Therefore so can ANY regular polygon with an even number of sides.

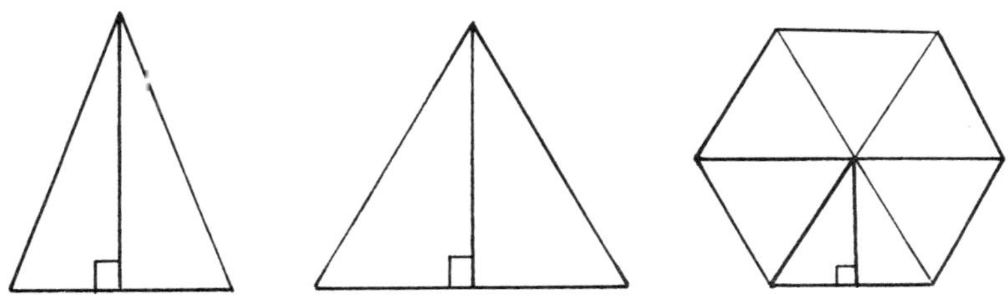

Angle of ELEVATION.

This is an angle between the horizontal and a point above the horizontal.

Think of looking horizontally and then lifting your eyes to look at the top of a cliff.

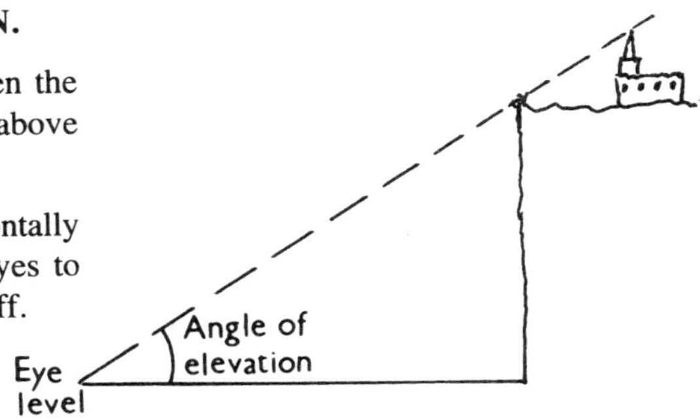

Angle of DEPRESSION.

This is the angle between the horizontal and a point below the horizontal.

Think of lying on top of the cliff and looking out horizontally. Then lower your eyes to look at a boat in the water.

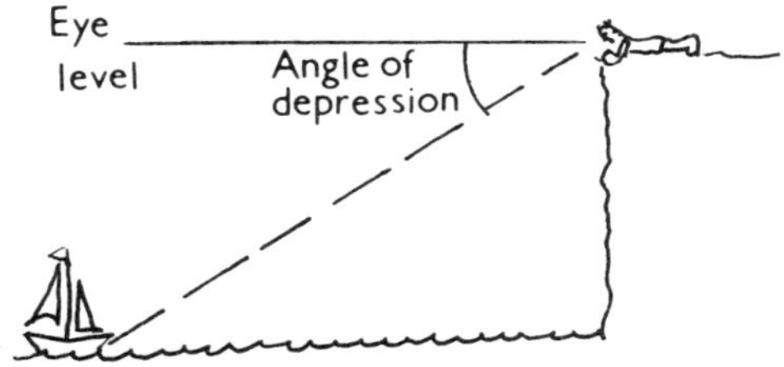

Bearings.

***RULE:* BEARING SHOULD ALWAYS BE TAKEN *CLOCKWISE* FROM THE NORTH LINE**

East is a 90° bearing.
South is a 180° bearing.
West is a 270° bearing.

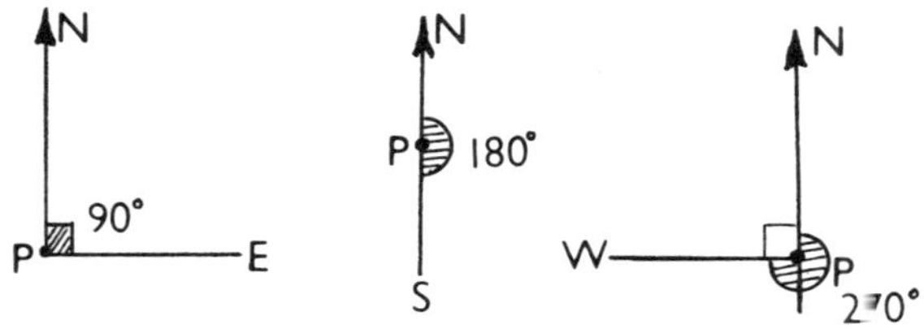

Sometimes bearings are given from the N or S lines towards E or W. Always read from the North unless otherwise instructed.

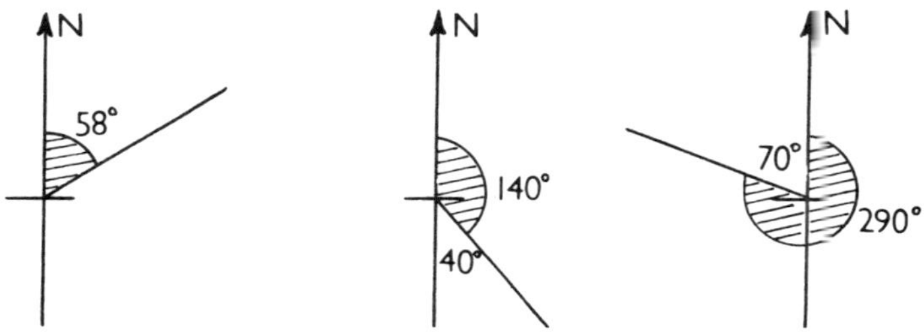

***RULE:* THE BEARING IS TAKEN AT THE PLACE *FROM* WHICH YOU ARE ASKED TO TAKE IT.**

RULE: ALWAYS DRAW IN YOUR NORTH LINE.

Example. What is the bearing of B <u>from</u> A.
What is the bearing of A from B?

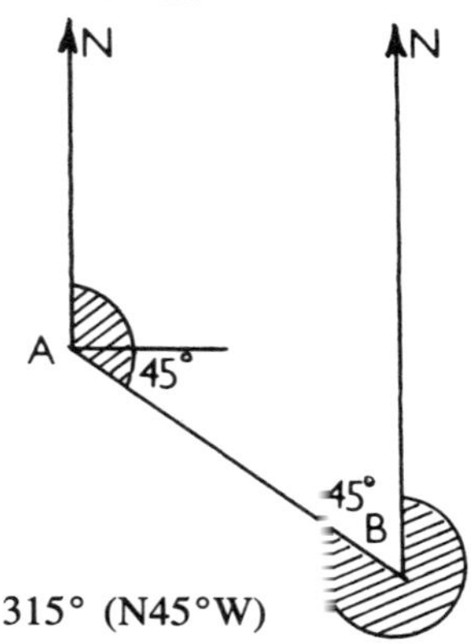

Bearing of B from A is 135°

Draw in the north line at B.
Bearing of A from B is the angle marked = 315° (N45°W)

Example. A ship sails 40 km due South from a port P to port Q. Then it sails on a course due East 60 km to port R. What is the bearing of port P from port R?

Draw the diagram.
First draw in our NORTH LINE at R.
Angle PRQ is calculated using Tan.

$$\text{Tan } a = \frac{O}{A} = \frac{40}{60} = 0.6666667$$

INV Tan gives 33.69°

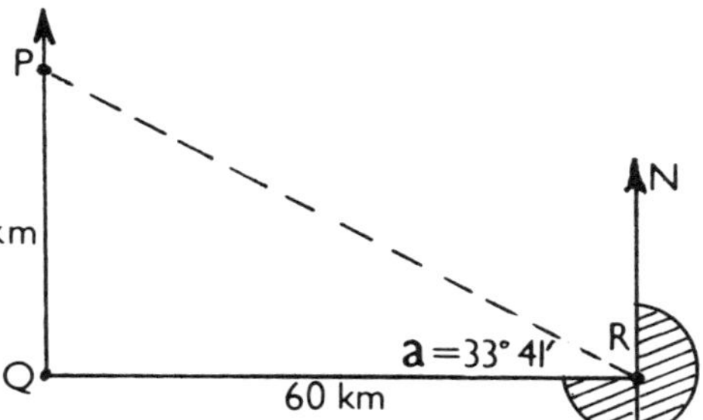

DMS gives 33° 41′ 24″. Size of angle *a*.

Bearing of P from R is 270° + 33° 41′ 24″. READ CLOCKWISE FROM N. LINE
= 303° 41 min. (to nearest minute).

Another way of working out angles and bearings is by SCALE DRAWING. Let 1 cm = 10 km and try it.

EXERCISE Q.

(1) The diagram below illustrates the path of a rocket. It reaches an altitude of 5 miles, then levels out to travel for 25 minutes at 1,200 km per hour. It descends at an angle of 24°. Find:

(a) angle SRV, the angle of ascent.
(b) the distance RS.
(c) the distance ST.
(d) the total horizontal distance travelled RVWU.

(2) A girl in a boat is 450 metres from the base of a cliff. The cliff top is 50 metres above sea level. When standing in the boat her eyes are 2 metres above sea level. She looks up to the top of the cliff. What is the angle of elevation from her eyes to the cliff top?

40. Construction and Scale Drawing

Given certain measurements of lines or angles we can construct a given figure.

TOOLS: (1) Pencil. (2) Ruler. (3) A pair of Compasses. (4) Protractor. (5) Rubber.

HINT.

We will be using the compasses to span lengths, so we will have to open them out along the ruler. To speed up this process, put the point of the compasses on the 0 line of the ruler and press it gently until you get a tiny hole on the line. When you have to measure off say 6 cm, all you do is put the point of the compasses in the hole and it stays there while you watch where the pencil end is being opened to. This increases speed and accuracy.

We use compasses to draw ARCS. Look at the arc below.

NOTICE that any point on the ARC is the same distance from the point A.

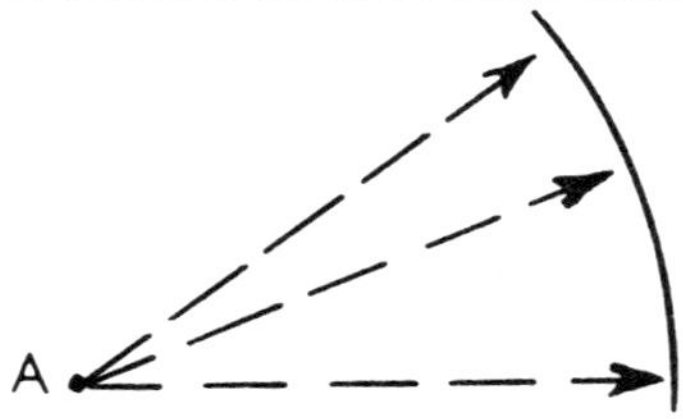

Example. Construct a triangle with sides 3 cm, 4 cm and 5 cm.

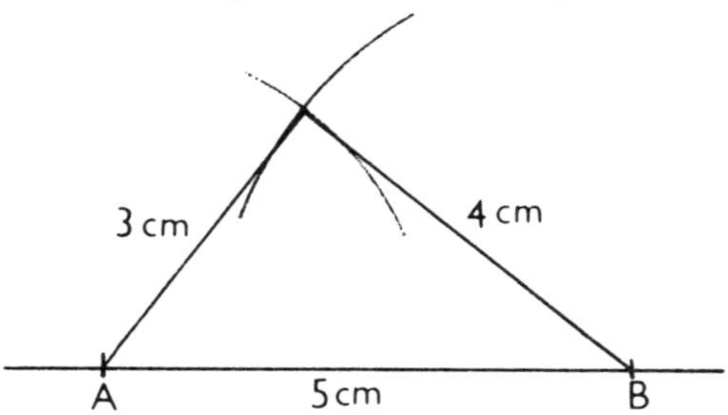

(1) We draw a line and mark two points A and B 5 cm apart.

NOTICE that the line is drawn longer than 5 cm to start with. A point on a line is easier to see than the end of a line.

(2) We open our compasses 3 cm and put the point on A.

(3) We draw an arc above the line.

NOTICE that it is good long arc.
Any point on that arc is 3 cm from A.

(4) We open our compasses to 4 cm, and put the point on B.

(5) Draw an arc above the line. Anywhere on this arc is 4 cm from B.

The point where the first arc is cut by the second arc is 3 cm from A and 4 cm from B.

BISECTING A LINE.

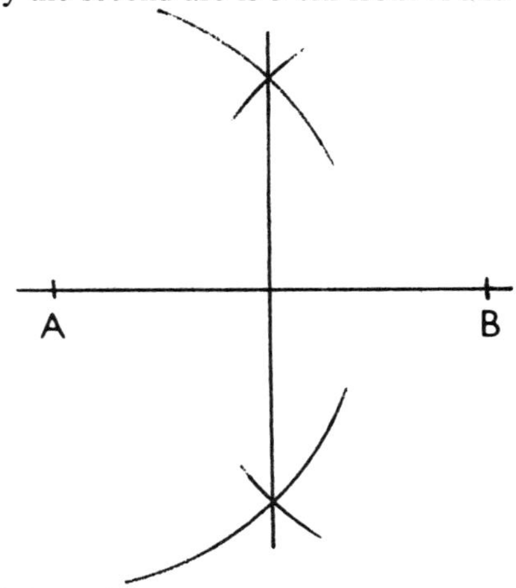

Given a line AB

Put compasses on A and open them wider than half of AB.

Draw arcs above and below AB.

Put compasses points on B.

Draw arcs above and below AB.

Join the two points where the arcs intersect.
This line is the PERPENDICULAR BISECTOR of AB.

CONSTRUCTING A RIGHT ANGLE.

Given line and point X. Construct a right angle at X.

First put compasses point on X and mark off points A and B with an arc at either side of X. A and B will be the same distance (equidistant) from X.

Compasses on A, arc above line.

Compasses on B, arc above line.

Join points where arcs cross to X. This line is PERPENDICULAR to line AB.

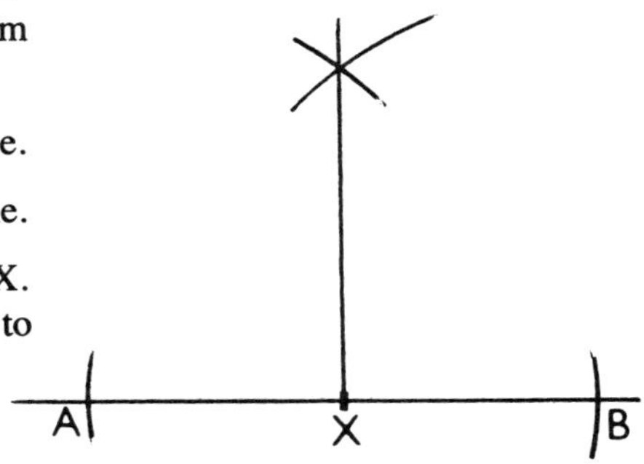

BISECTING AN ANGLE.

Given angle at A.
Put compasses point on A
and arc through both lines.

Mark the intersections B and C.
Put compasses point on B
and draw arc X.

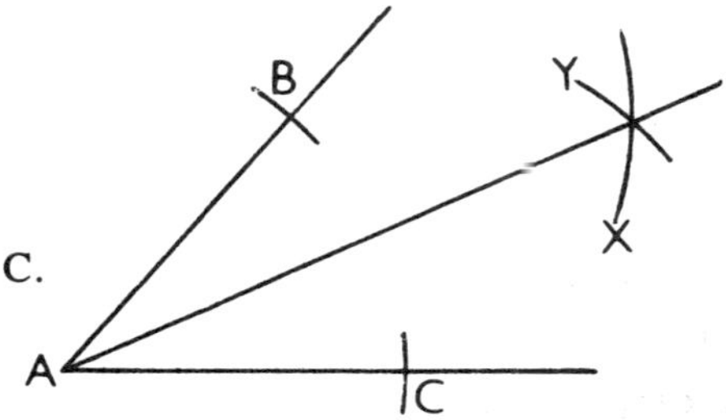

Put compasses on C and draw arc Y.

Connect intersection of X and Y to A.

45° ANGLE.

By bisecting a right angle we can obtain an angle of 45° without using a protractor.

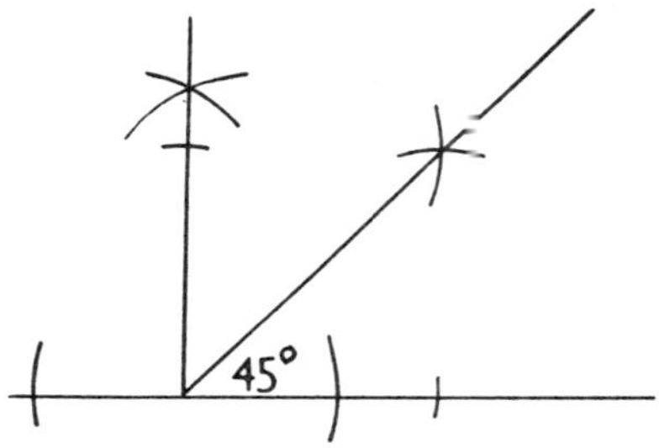

60° ANGLE.

We can construct an angle of 60° without using a protractor by constructing an EQUILATERAL triangle.

Draw line AB.

Put compasses point on A.
Arc through line at B.
Arc above the line.

Put compasses point on B.
Arc through first arc at C.

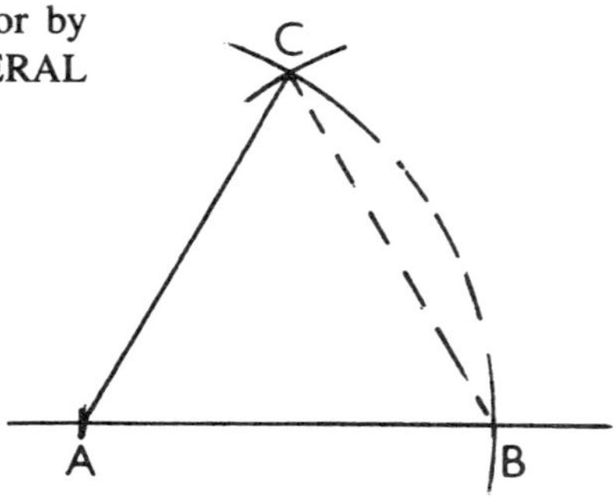

The point where the arcs cross is equidistant from A and B.
AB = CA = CB.

Join the points and all the angles are 60°.

You can see that by using ARCS we can *PLOT* the same distance anywhere along it.

***RULE:* DON'T BE SHY WITH YOUR ARCS.**

A good sweeping arc is better than a short one. An intended intersecting arc can fall short of a short arc.

OTHER CONSTRUCTIONS

Using the PROTRACTOR we can construct triangles given 2 sides and an INCLUDED angle or 2 angles and 1 side in between them.

Example. Draw triangle ABC where AB = 4.5 cm, BC = 2 cm and angle ABC = 34°

Draw line BC.
Measure 34° at B.
Draw in line along BA.
Use compasses opened to 4.5 cm on B and arc across BA.
Join A and C.

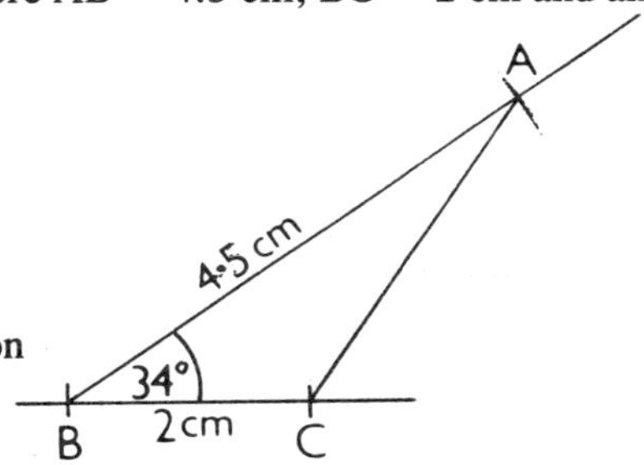

Example. Draw triangle ABC where BC is 4 cm, angle ABC = 30°, and angle ACB = 45°

Draw line and mark off B and C 4 cm apart.
Measure angle at B and angle at C and draw in the lines until they intersect at A.

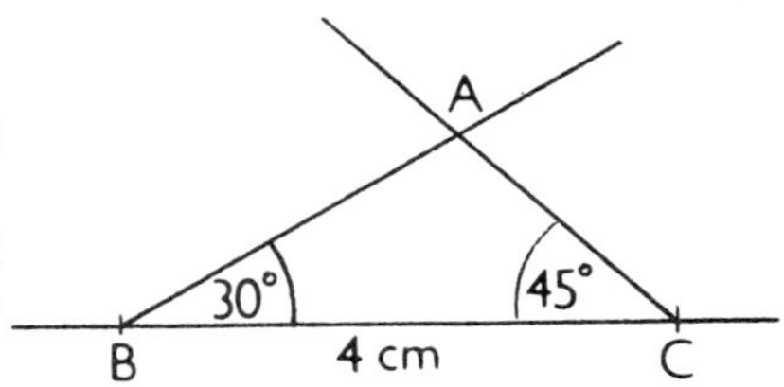

HINT: Draw a longer line and mark off points B and C. Then when we measure angle ABC, we can see clearly the point we are measuring from, much more easily than the end of a line.
We can also square up the lines on the protractor more easily.

SCALE DRAWING

Scale drawings are done in exactly the same way as the above constructions except that we scale down the measurements to fit on the paper. Measurements are then made with ruler or protractor.

Example. A boat B sails east of port P for 35 km and then turns due south. What is the bearing of the port P from the boat B when it has travelled 25 km south?

Scale 1 cm = 10 km

Bearing = 360° − 54° = 306°

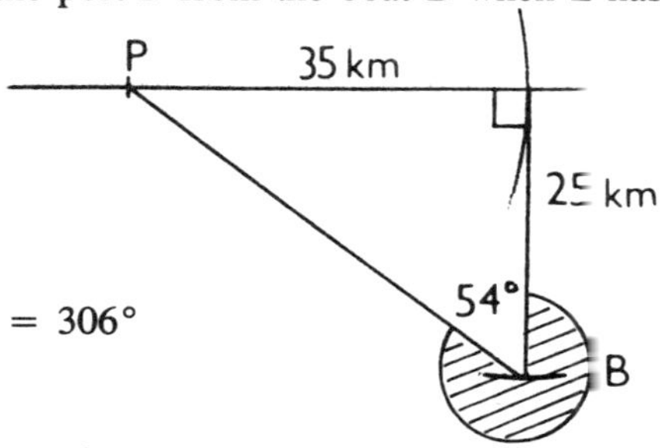

When the construction is complete, REMEMBER to convert your measured cm back to km or metres or whatever.
ANGLES will remain the same of course.

OTHER USES OF COMPASSES

Tracing rotations

***RULE:* IF AN IMAGE IS ROTATED ABOUT A POINT THEN EACH POINT ON THAT IMAGE IS ROTATED SEPARATELY.**

Example. Rotate the image P anti clockwise about point A onto B through an angle of 90°

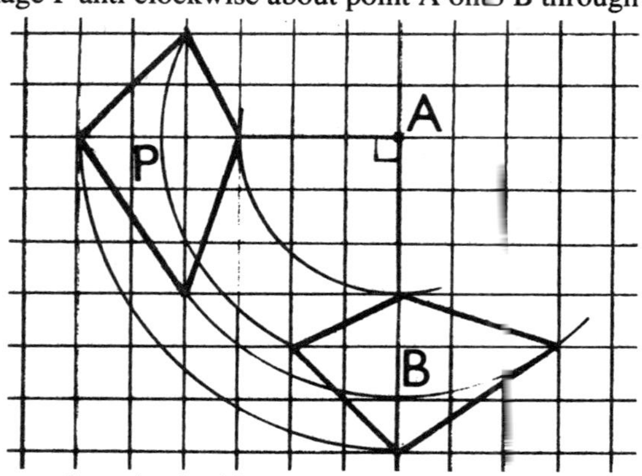

There can be little chance of a mistake if you trace each point.

Conversely, given 2 images, we can find THE CENTRE OF ROTATION and check it by using our compasses.

NOTICE:

***RULE:* WITH 90° ROTATIONS *EACH* POINT ON THE IMAGE WILL BE EXACTLY AT RIGHT ANGLES TO THE SIMILAR POINT ON THE ROTATED IMAGE.**

41. Algebraic Graphs

AXES.

The axes of all algebraic graphs are the X and Y AXES.

***RULE:* THE X AXIS IS ALWAYS HORIZONTAL.**
THE Y AXIS IS VERTICAL.

COORDINATES.

Points on graphs are called Cartesian Coordinates. They are a grid reference.

***RULE:* ALWAYS READ THE X COORDINATE FIRST.**

The point A on the graph is (4,3). That means you read ON the X axis 4, and UP the Y axis 3.

They can be at the minus side of X like point B (−1,3), or on the minus side of Y, like point C (2,−3) or both like D (−5,−2)

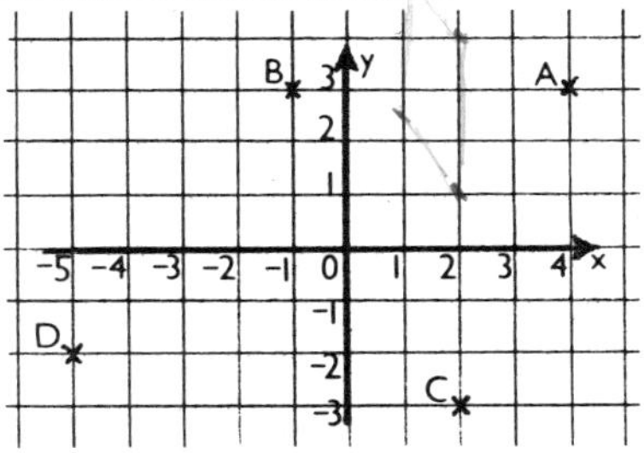

TABLES OF VALUES.

All graphs can have a table of values. In a formula t = $2x$, we can put in our own values for x and get a value for t.

When x	=	0	1	2	3	4	5	6
t = $2x$	=	0	2	4	6	8	10	12

We can plot these on a graph.

See when x = 1, t = 2
and when x = 2, t = 4 etc.

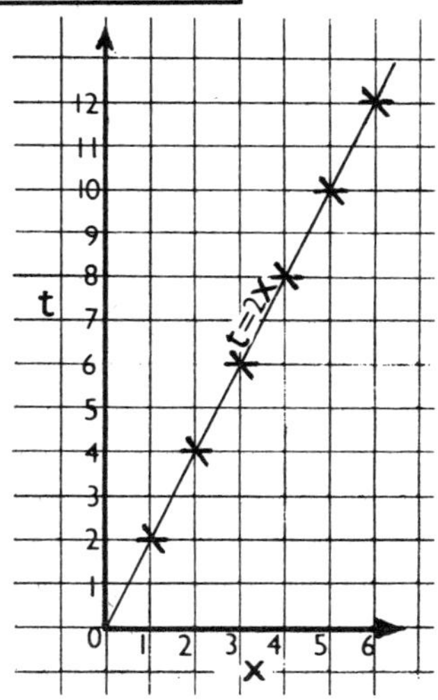

EQUATION OF A STRAIGHT LINE

Straight lines on a graph can be expressed as an equation.

$y = 3x + 2$ is an equation of a straight line.

We can plot it

When	x	=	1	2	3	4
	$3x$	=	3	6	9	12
	2	=	2	2	2	2
So	y	=	5	8	11	14

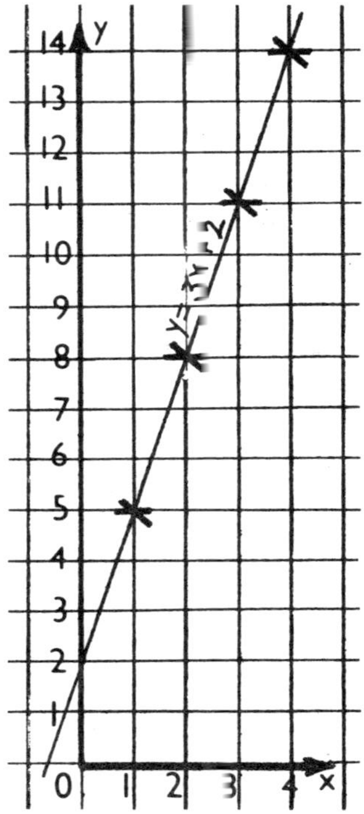

***RULE:* THE EQUATION OF A STRAIGHT LINE IS IN THE FORM**

$$\mathbf{y = m}x \mathbf{+ c}$$

Where **m** is the GRADIENT

and **c** is the point where the line cuts the Y axis.

(Called the y intercept)

So in $y = 3x + 2$

3 is the gradient

2 is the point on the Y axis where the line cuts the Y axis.

REMEMBER ALWAYS EXTEND THE LINES BEYOND THE POINTS MARKED.

REMEMBER ALWAYS label the X and Y axes.

REMEMBER ALWAYS label the line with its equation.

42. Gradient

Think of a hill going up 1 in 4
As with any ratio we can write this $\frac{1}{4}$

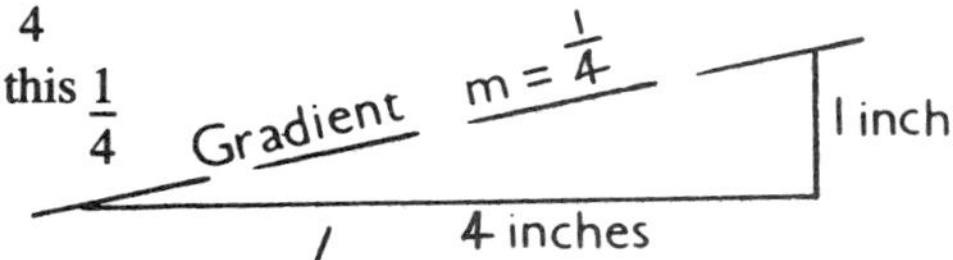

We climb 1 inch vertically for every 4 inches we travel horizontally.

It is exactly the same on graphs.

A gradient of 3 is $\frac{3}{1}$ or 3 in 1

That's 3 UP for 1 ON.

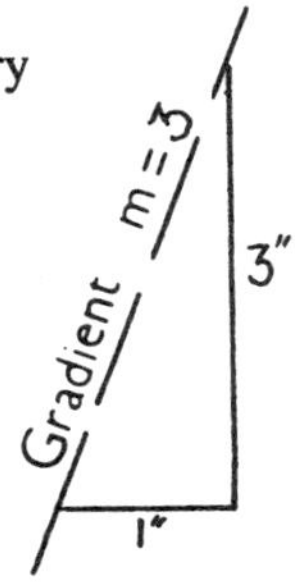

A gradient of $\frac{1}{2}$ (1 in 2) is 1 UP for 2 ON.

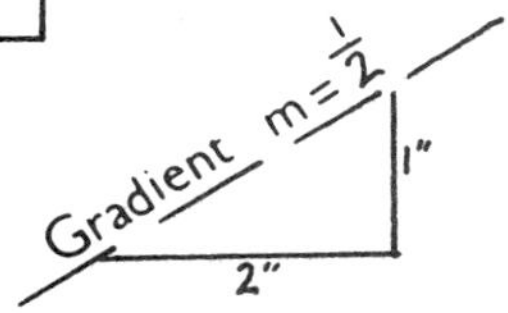

RULE:* THE TOP NUMBER IS THE VERTICAL CLIMB. *UP
THE BOTTOM NUMBER IS HORIZONTAL TRAVEL. *ON*

Example. Plot the graph of the line $y = 3x + 1$

We can either put in a table of values
When

x	=	0	1	2	3	4
$3x$	=	0	3	6	9	12
1	=	1	1	1	1	1
y	=	1	4	7	10	13

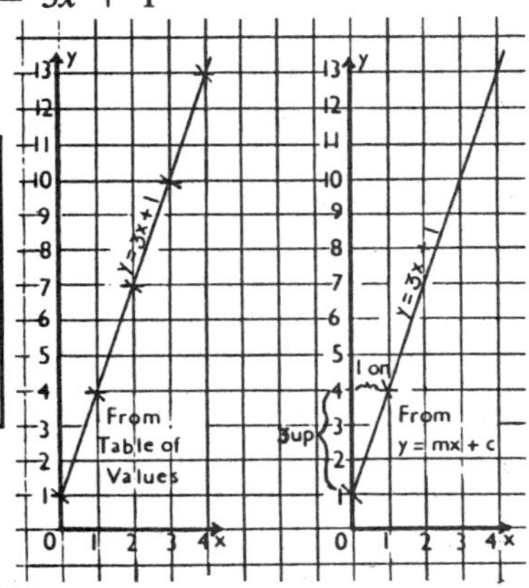

OR we can say the gradient **m** is 3

It cuts the Y axis at **c** (0,1)

From the point (0,1) go 3 UP and 1 ON.

Mark the spot.

Join it to point (0,1)

IT IS THE SAME LINE.

Example. In the graph, give the equation of line AB.

The Gradient **m** is $\frac{3}{1}$ 3 UP and 1 ON.

It cuts the Y axis at -2 which is **c**.

Equation of the line AB is $y = 3x - 2$.

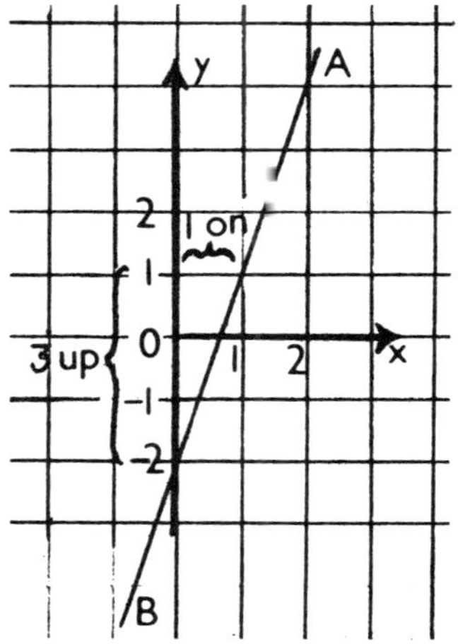

NOTICE **c** can be a minus.

So can **m**.

RULE: **A POSITIVE GRADIENT ALWAYS SLOPES DOWN FROM RIGHT TO LEFT**
MINUS GRADIENT MEANS THE LINE SLOPES DOWN FROM LEFT TO RIGHT

Example. Plot the equation of the line $2y + x = 4$ and mark it AB.

This has to be put in terms of a single y.

$$2y = -x + 4$$

Divide by 2 and get $y = -\frac{1}{2}x + 2$

Here **c** $= 2$

m $= -\frac{1}{2}$

So the line slopes down from left to right.

Go to point (0,2) on the Y axis (which is **c**)

Go UP 1 and ON 2 **TO THE LEFT**

Mark the point.

Join that point to point (0,2).

That is line AB.

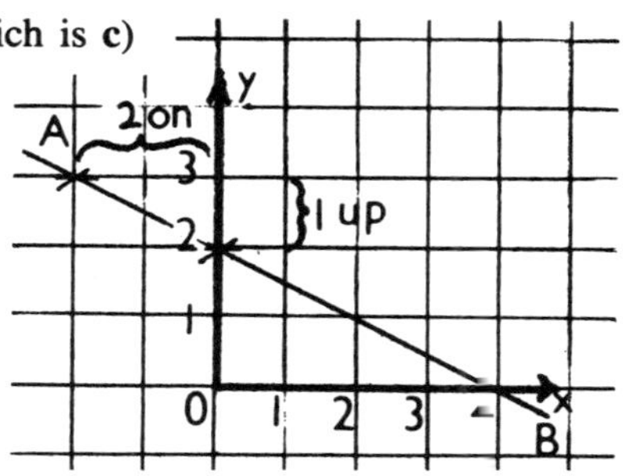

EQUATIONS OF OTHER LINES.

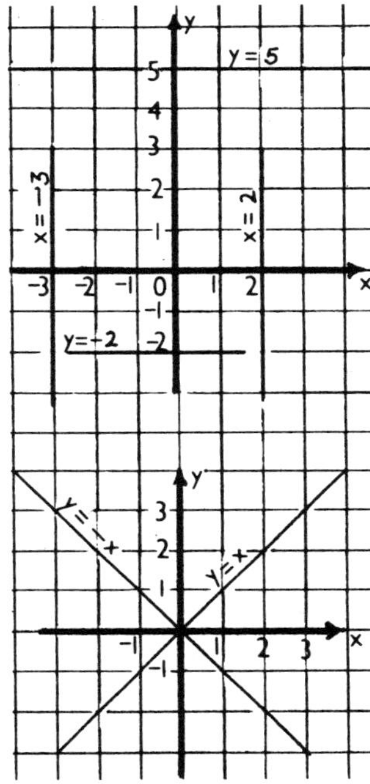

x = 2.
Where a line runs parallel to the Y axis say at the point (2,0) then $x = 2$ all along that line.

y = 5.
Where a line runs parallel the X axis say at point (0,5) then $y = 5$ all the way along that line.

y = x.
Where x and y coordinates are the same (1,1) and (2,2) and (3,3) etc. then the equation of the line is $y = x$.

y = −x.
Where the coordinates of a line fall on (−1,1) and (−2,2) etc. then the equation of the line is $y = -x$.

There are many more.
Draw in a few of your own.

EXERCISE R.

(1) Construct a triangle whose base angles are 38°, and 58°. The base is 5 cm long, how long are the other two sides?

(2) Draw an angle of 50° using your protractor. Then bisect it with compasses only. Check the size of the angle with your protractor.

(3) In the equation of a straight line $y = 3x + 2$,

(a) what are the coordinates of the point where it cuts the Y axis?

(b) what is the gradient?

(c) draw the graph of the line on squared paper and label it.

(4) On the same graph draw the line $x = -1$.
Where is the line of symmetry between the line $y = -1$ and the line $y = 5$? Draw it in and label it with its equation.

(5) What are the coordinates of the points where the line $y = 4x + 2$ crosses the axes?

(6) Where does the line $y = 3x - 3$ cross the axes?

(7) Where does the line $y = 5$ cross the axes?

(8) What is the gradient of the line that passes through the points (1,1) and (4,3)?

43. The Exam

Exam technique is a qualification on its own. Learning how to take *ANY* exam means that you can apply what you have learned to other situations.

(1) **A JOB WELL DONE IS NOT ALWAYS WORTH DOING.**

You can spend hours or days perfecting something, only to find it never was going to turn up in the exam. So get your syllabus NOW. The first thing to learn is to KNOW YOUR TERMS OF REFERENCE.

DO ONLY WHAT YOU ARE ASKED TO DO.

In other words, know exactly what will NOT come up in the exam. What is not in the syllabus will not be in the exam. What *is* in the syllabus MAY be in the exam.
The old saying 'A JOB WORTH DOING IS WORTH DOING WELL' is an excellent guide to learning any topic. But don't use precious time learning something that you will not be examined on.

(2) The above also applies *IN* the exam. Don't give the examiners answers to questions they haven't asked even if those topics are in the syllabus. Doing so is a complete waste of time and no marks will be gained.

(3) Get the past exam papers or specimen papers. See what subjects the most popular questions are about. Some things ALWAYS come up in an exam. Guarantee yourself maximum marks on those subjects by making sure you know them backwards!

(4) Learn all the topics to which THE RULE OF X can be applied. Most of them will come up in the exam papers.

(5) In the exam READ THE QUESTION.
ANSWER IT.
READ THE QUESTION AGAIN and make sure you have given them what they asked for.

(6) If there is a question you have difficulty in understanding, then LEAVE IT and go onto the next one. Go back to the question you have missed when you have answered ALL the other questions. That way if you run out of time you will not lose marks on questions you could have answered but never got time to answer because you were bogged down with one tough question.

(7) Take all your tools, and several pens and pencils. Don't forget your Rubber.

(8) For a major exam I suggest you put new batteries in your calculator to ensure they don't run out at the time you rely on it the most.
Get to know YOUR calculator. Calculators are not all the same. Some calculators, for example, may require you to press INV before °'''.
So get your calculator and learn how all its functions are operated.

(9) Finally, DON'T PANIC! Confidence helps logical thought. Confidence can be built up at home by completing past exam papers in the time allocated. Local Libraries will have volumes of past exam papers in the Reference Section.

FINAL WORD. LEARN BY YOUR MISTAKES

If you get a question wrong then it means you need to study that topic some more. Learn the topic until you realise where you went wrong. Practice makes perfect.

44. Examination Boards

AEB	Associated Examining Board, Wellington House, Aldershot, Hampshire GU11 1BQ.
UCLES	University of Cambridge Local Examinations Syndicate, Syndicate Buildings, 17 Harvey Road, Cambridge CB1 2EU.
JMB	Joint Matriculation Board, Manchester M15 6EU. Publications from John Sherratt & Sons Ltd. 78 Park Road, Altrincham, Cheshire.
LONDON	University Entrance and Schools Examination Council University of London, 66-72 Gower Street, London WC1H 0AP. Publications Office 52, Gordon Square, London WC1H 7BX.
OXFORD	Oxford Local Examinations. Delegacy of Local Examinations, Ewert Place, Summertown, Oxford OX2 7BX.
OXFORD & CAMBRIDGE	Oxford and Cambridge Schools Examination Board. 10 Trumpington Street, Cambridge & Elasfield Way Oxford.
SUJB	Southern Universities Joint Board for School Examinations Cotham Road, Bristol BS6 6DD.
WJEC	Welsh Joint Education Committee, 245 Western Avenue, Cardiff CF5 2YX.
SEB	Scottish Examination Board Ironmills Road, Dalkeith, Midlothian EH22 1BR. Publications from: Robert Gibson & Sons Ltd. 17 Fitzroy Place, Glasgow G3 7SF.
NIGCE	Northern Ireland General Certificate of Education Examination Board, Beechill House, 42 Beechill Road, Belfast BT8 4RS.

45. Answers to Exercises

(Page 8).

A. 1. (1) 0.127 (2) 35.69 (3) 0.07 (4) 0.017
(5) 0.004 (6) 70 (7) 1600 (8) 10 (9) 0.79 (10) 8400

(Page 8).

A. 2. (1) 11, 19, 31 and 41.
15 is divisable by 3 and 5 and 81 is divisible by 9 and 3
(2) 31, 37, 41, 43, 47.

(Page 8).

A. 3. (1) 2, 3, and 5 (2) 3 and *a* since 3 × *a* is 3a

(Page 9).

A. 4. (1) 2 and 7 (2) 2 and 7

(Page 9).

A. 5. (1) 102 (2) 70 (3) 2 × 2 × 2 ×3

(Page 9).

A. 6. (1) 49 (2) 225 (3) 15^2

(Page 10).

A. 7. (1) 19 (2) 169 (3) 27.92848 (4) 25.6

(Page 11).

A. 8. (1) 2 × 2 × 2 × 2 × 3 × 3 × 7 (2) $2^4 \times 3^2 \times 7$ (4) 7

(Page 14).

B. 1. (1) $4\frac{3}{8}$ (2) $\frac{3}{8}$ (3) $\frac{1}{8}$ (4) $1\frac{3}{4}$ (5) $4\frac{1}{5}$

(Page 17).

B. 2. (1) 0.8 (2) $\frac{19}{25}$ (3) 0.457 (4) 34.56 (5) 34.6 (6) 34.56
(7) 3450 (8) 3460 (9) 15 (10) 18 (11) 34 (12) $\frac{3}{7}$
(13) 38 metres (14) 2 hr 47 min (15) 1 hr 49 min.

(Page 44).

C. (1) £149.50 (2) £11.70 (3) £1269 (4) £288
(5) £1.62 (6) £1.40 (7) 24% (8) 22% & 78%
(9) £150 (10) £5188.80 (11) (a) 29.5 mpg; (b) 5.70 gal.;
(c) 25.9 litres; (d) 37.8p (12) (a) 875 grammes; (b) 10 kg
(13) 50 mph (14) (a) 30%; (b) £3.92; (c) 84 pence
(15) 24,000 cm^3 (16) (a) 15.5 litres per minute; (b) 3 min.

(17) 35,280 lire (18) (a) 3975 francs; (b) £4.36; (c) £6.20 (19) 1.2 cm (20) 2.4 cm (21) (a) 1:4 (b) 11:1; (c) 1:5; (d) 2:7; (e) 10p:3p (22) 8% (23) £6352.50 (24) 126° (25) (a) 4.5 cm; (b) 20 cm (26) 6 hours (27) 5 hrs for 2 kg (28) (a) 84 days (b) 192 days (29) Only 9 cuts needed, 18 minutes. (30) (a) 1 min 36 sec; (b) 32 sec. (31) (a) £396 and £264; (b) £330, £220, £110. (32) 127.

(Page 49).

D. (1) (a) X = 3 (b) X = 6 (c) X = 10 (2) 3a + 13b
(3) (a) 3X − 3Y = 7 (b) a = 2b (c) a = −b OR b = −a
(4) (a) X = 3Y −2 (b) X = 5 − 4Y (c) X = 3Y + 2
(5) $R = \frac{100I}{PT}$ (6) $f = \frac{a}{t^2}$ (7) $S = \frac{9F - 2}{27}$
(8) $P = \frac{X(2L - 1)}{3Y}$ (9) $S = \frac{F - 4}{9}$ (10) X = Y − 2

(Page 50).

E. (1) (a) $6a^2b$; (b) $6a^3b$; (c) $2a^3b^3$
(2) 2 into 4 goes twice, a into a goes a giving 2a (3) $\frac{3}{a^4}$ (4) ab^{-2} is $a \times b^{-2}$ is $\frac{a}{b^2}$ (5) a^3b

(Page 52).

F. (1) Z = 11 (2) $b = \frac{3}{4}$ (3) (a) 2X pence (b) (2X + 10) pence (c) 14 pence each.

(Page 58).

G. (1) 7 years, £102 (2) 7% (3) £400 (4) £800 (5) 5 years, £106.29. Enter 60 and see how many times you have to multiply by 1.10 (6) (a) 2600×1.13^5 = £4790.33; (b) (i) 2600 + 13% interest is £2938. Take away 800 at end of year leaving balance of 2138. 2138 × 1.13 (that is + 13%) is £2415.94. Take away 800 leaves £1615.94 owed at beginning 3rd yr. (ii) 1615.94 × 1.13 = 1826.01. Take away 800 and we have 1026.01 at beginning of 4th yr. 1026.01 × 1.13 = 1159.39. Take away 800 and we are left with £359.39 at beginning of 5th yr. 359.39 × 1.13 = £406.11 to pay off loan at end of 5 yrs. TOTAL paid to bank was (4 × 800) + 406.11 = £3,606.11p.
(7) $1 - \frac{1}{2} - \frac{1}{4} - \frac{1}{8} - \frac{1}{16} - \frac{1}{32} = \frac{1}{32}$ or 0.03125
(8) 700×0.85^4 = £365.40

(Page 61).

H. (1) £125.50 (2) (a) £1144 rent + £350 gas + £363 rates (b) total is £1857

(3) (a) £204 × 52 = 10,608

less 3,450

7,158 taxable at 24%

tax is 1,717.92

10,608 − 1,717.92 is 8890.08 after tax divide by 52 to get £170.96 per week net wage (b) 23,570 − 4,670 leaves 18,900 taxable. 1st 16,700 @ 24% is £4,008 tax, remainder 2,200 @ 40% = 880 tax. Total tax £4,888 from £23,570 leaves net salary £18,682

(4) (a) 487 units; (b) 603.88 therms; (c) £240.34; (d) £250.14 (5) (a) £6.60 (b) £176 (c) £209 (d) 12 hrs.

(Page 66).

I. (1) For 6 exams 63 × 6 = 378 total marks

For 7 exams 68 × 7 = 476 total marks

Difference between 6 exams and 7 exams is 476 − 378 = 98 marks gained in 7th exam.

(2) (a) $7X + 14$ (b) $X^2 + 2X - 8$ (c) $X^2 + 8X + 16$; (d) $X^2 - 8X + 16$ (3) 15 days (Did you use the rule of X?)

(4) (a) $12 \times (5 - 3) = 24$ (b) $(11 \times 2) - 6 = 16$

(c) $[(7 + 4) \times 3] + 2 = 35$ (5) (a) £4 (b) £7

(Page 67).

J. (a) $X = 5$ (b) $X = 12$ (c) $X = 10$ (d) $X = \frac{1}{8}$ (e) $X = -1$

(Page 70).

K. (1) 2.31×10^8 (2) 6.5×10^{-5} (3) 0.00045 (4) (a) $X < 6\frac{1}{7}$; (b) $X = 6$

(5) $8.75 \times 10^{-2} = 0.0875$ seconds to travel 105 feet.

105 ft in 0.0875 sec gives $\frac{105 \times 1}{0.0875}$

X ft in 1 sec

Velocity = 1200 feet per second

(Page 79).

L. (1) (a) $2X(5X - 1)$ (b) $(a - 3b)(2 - 4a)$

(2) (a) $\frac{1}{3} \times \frac{1}{3} \times \frac{1}{3} = \frac{1}{27}$ (b) $\frac{1}{3} \times \frac{1}{3} = \frac{1}{9}$ (c) $\frac{1}{8} + \frac{1}{8} + \frac{1}{8} = \frac{3}{8}$

(d) equal chance of two boys or two girls $= \frac{1}{2}$

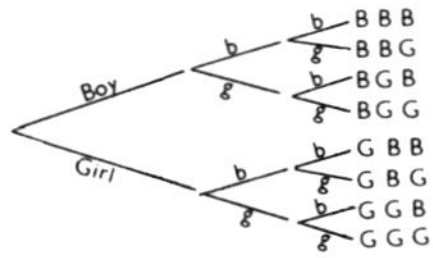

(3) (a) $\frac{5}{7}$ (b) $\frac{1}{7}$; (c) $\frac{5}{7} \times \frac{2}{3} = \frac{10}{21}$; (d) $\frac{1}{7} + \frac{1}{7} = \frac{2}{7}$

(Page 85).

M. (1) 8 cm (2) 36 cm (3) 32.450 km (4) 71° each (5) a = 53°, b = 53°, c = 93°.

(Page 90).

N. (1) a = 90°, b = 90°, c = 23°, d = 113° (2) a = 47.5°, b = 47.5°, c = 42.5°, CX = 3 cm.

(Page 100).

O. (1)

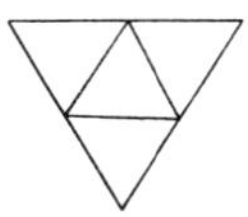

(2) 390 cm³ (3) 108°
(4) Rhombus has 2 pairs of parallel sides. Trapezium has one pair.
(5) 8 cm.
(6) Points of contact add up to 360°
(7) (a) 60 cm²; (b) 4.25 m.
(c) A 5 cm cube is 125 cm³
Use RULE OF X: 125 cm³ = 0.925 kg. That gives 188.7 kg.
25,500 cm³ = X;
(d) 7.4 grams. (8) 144°

(Page 104).

P. (1) (a) Yes. (b) No. (c) Yes. (2) scale factor 3
(3) (a) A reflection on the line $y = x$ (b) 180° rotation.
(c) Vector translation $\begin{pmatrix} -8 \\ 0 \end{pmatrix}$

(Page 111).

Q. (1) (a) 22° 37′ 12″ (b) 13 miles (5, 12, 13 triangle)
(c) 500 miles (d) 523.23 miles (2) 6° 5′ 18″

(Page 121).

R. (1) 3.1 cm and 4.2 cm. (3) (a) (0,2) (b) 3 (4) line y = 2.
(5) (0,2) and (−0.5, 0). (6) (0, −3) and (1,0). (7) (0,5) only.
(8) $m = \frac{2}{3}$